Decentralizing
Knowledges

Leandro Rodriguez Medina and
Sandra Harding, Editors

Decentralizing Knowledges

ESSAYS ON DISTRIBUTED AGENCY

DUKE UNIVERSITY PRESS *Durham and London* 2025

Project Editor: Livia Tenzer
Designed by MaTThew Tauch
Typeset in Arno Pro by Westchester Publishing Services

Library of Congress Cataloging-in-Publication Data
Names: Rodriguez Medina, Leandro, editor. | Harding, Sandra G.,
editor.
Title: Decentralizing knowledges : essays on distributed agency /
Leandro Rodriguez Medina and Sandra Harding, eds. Description:
Durham : Duke University Press, 2025. |
Includes bibliographical references and index.
Identifiers: LCCN 2024044711 (print)
LCCN 2024044712 (ebook)
ISBN 9781478031796 (paperback)
ISBN 9781478028550 (hardcover)
ISBN 9781478060772 (ebook)
ISBN 9781478094289 (ebook other)
Subjects: LCSH: Knowledge, Theory of. | Knowledge, Sociology of. |
Feminist theory. | Feminist ethics.
Classification: LCC HM651 .D435 2025 (print) | LCC HM651 (ebook) |
DDC 121—dc23/eng/20250118
LC record available at hTTps://lccn.loc.gov/2024044711
LC ebook record available at hTTps://lccn.loc.gov/2024044712

Cover art: Sarah Sze, *Shorter than the Day*, 2020. Powder-
coated aluminum and steel, 48×30×30 ft. © Sarah Sze.
Photo by Nicholas Knight. Commissioned by LaGuardia
Gateway Partners in partnership with Public Art Fund.
Courtesy of the artist and Gagosian.

This book is freely available in an open access edition thanks to TOME
(Toward an Open Monograph Ecosystem)—a collaboration of the
Association of American Universities, the Association of University
Presses, and the Association of Research Libraries—and the gener-
ous support of Arcadia, a charitable fund of Lisbet Rausing and Peter
Baldwin, and the UCLA library. Learn more at the TOME website,
availableat:openmonographs .org.

Contents

III   Creating Alternative Spaces

Acknowledgments

We have tried to make the very task of writing this book an exercise in epistemic decentralizing. After a series of virtual meetings with the authors, we loosely defined the central concept to accommodate their research, perspectives, and critiques. Each one entered into a dialogue that yielded interesting and unexpected insights, in addition to teaching us the levels of the phenomenon, around which we have structured the book. We would like to thank each and every participant for their generosity, their patience, and their willingness to contribute innovative and useful ideas to a debate that is ultimately about the universal possibility of producing knowledge, that is, about epistemic justice.

Kenneth Wissoker and Ryan Kendall of Duke University Press have provided pertinent questions, insights that enhanced the manuscript, and confidence in this project from its inception. With them, through multiple exchanges, we have been able to advance our own thoughts on epistemic decentralizing, hoping that readers will now encounter a deeper, more thoughtful, and enlightening elaboration. The anonymous reviewers of the book proposal pointed out clear weaknesses and suggested fruitful paths along which we editors and authors have circulated. Thus, working with Duke University Press has been a rewarding challenge and an undeniable learning experience.

A key contribution to the decentralizing of the knowledge contained in this volume has been the possibility of making it open access. We would like to acknowledge the generous support of the UCLA Library through Sharon Farb and the TOME (Toward an Open Monograph Ecosystem) initiative that made this a reality.

In recent years, the editors have had the opportunity to discuss some of the ideas that have materialized here. Leandro Rodriguez Medina would like to thank Rigas Arvanitis, Fernanda Beigel, Renato Bermúdez Dini, Ronald Cancino, Rosalba Casas, Michelle Chauvet, Kathleen Cruz Gutiérrez, David Dumoulin, Kim Fortun,

César Guzmán Tovar, Wiebke Keim, Mina Kleiche-Dray, Susanne Koch, Pablo Kreimer, Emanuel Kulczycki, Alberto López Cuenca, David Mills, Jaimie Morse, Alis Oancea, Angela Okune, Anne Pollock, Andrés Seguel, Ari Sitas, and Hebe Vessuri. Sandra Harding thanks her numerous colleagues and students at both the University of Delaware and the University of California, Los Angeles for their insights and comments on these issues through the years. We are both deeply grateful to Mamta Jha for indexing this volume, and to Luisa Grijalva, not only for the fantastic editorial work of getting this manuscript ready for submission to the publisher, but also for years of rigorous and loving work at *Tapuya: Latin American Science, Technology and Society*. This book is also the product of this serendipitous and powerful assemblage that has been taking shape for eight years.

Epistemic Decentralizing

Distributed Agency in a Context
of Knowledge Asymmetries

With the critical phase of the COVID-19 pandemic over, almost no one seems to doubt that the world has changed and that the implications of this global phenomenon are yet to be seen and understood. It is a world of environmental, economic, political, and epistemic uncertainties. Climate change that threatens life on the planet, inflation rates that push millions into poverty, and a European war that threatens to become global and perhaps nuclear characterize an unstable environment. Social groups with divergent interests, from terraplanists to artificial intelligence techno-optimists through antivaccine and terrorist groups, have understood that every sociopolitical struggle is also epistemic and have turned, with better or worse results, to producing knowledge in pursuit of their objectives. As in previous moments of history, generalized uncertainty, paradoxically, appears as a consequence of the proliferation of agents demanding their epistemic status (Blair 2011). Therefore, in this environment, knowledge no longer emerges as a guarantee of solutions but rather as a terrain for disputes.

During a good part of modernity, as Latour (1987) and Burke (2000) have described, some European powers have accumulated information about the rest of the world, collecting specimens, maps, diagrams, and other forms of evidence. By so doing, their scientists and engineers became able to "intensify the race of evidence. . . . The tiny number of scientists is more than compensated by the great number of resources they are able to gather" (Latour 1987, 232). By gathering and analyzing, these sites became centers of calculation (Jöns 2011), and the knowledge emanating therefrom contributed, simultaneously, to the consolidation of the science that was produced and of the power of the empires that supported it. With differences, this centralization of knowledge continues to this day, but knowledge itself, once presented as a way to evade the demons of doubt and to exalt human reason, has been called into question in the beginning of the twenty-first century (Nieto Olarte 2019). Perhaps driven by the once-radical assertion that knowledge is a social construction, many actors have come to understand that in transepistemic arenas the laws of nature are not discussed innocently and disinterestedly, but instead alternative worlds are constructed (Law and Urry 2004). Thus, controversies about knowledge have become deep disputes over sociopolitical orderings (Law 1992), and vice versa, as was shown by the race for anti-COVID vaccines between corporations and research centers and between Western and non-Western powers (Meghji and Niang 2022).

Thus, this geopolitical context looks like a hotbed of new actors competing to impose their truth. It appears as a game of economic and political power that mixes propaganda and fake news, and prestigious institutions and lax social networks mediated by technologies. It can also be thought of as an environment in which we have finally visualized cracks in the monolithic edifice of legitimized knowledge—generally patriarchal, capitalist, and colonial (Lobo 2022). Differently placed, taking advantage of the breaks, ruptures, and liminal spaces provided by an environment of very high uncertainty, actors seem to have decided that it is a good time to make their claims heard and to build alternative forms of social organization (Masaka 2021; Ude 2022). Indignados, Occupy Wall Street, #MeToo, Black Lives Matter, Anonymous, Just Stop Oil, and cryptocurrency enthusiasm are just a few examples of groups that have put on the table an alternative and urgent agenda—decentering. At the same time, they strive for new infrastructures and practices that are political, social, economic, and cultural, and that allow them to feed that agenda—decentralizing. Unlike other social movements throughout history, which have focused fundamentally on the conquest of political power, these understand that one of the most important areas of dispute, if not the most important, is knowledge (de Sousa Santos 2018; Hess 2009; Frickel et al. 2010). For this reason, instead of engaging in the well-known and contradictory relations with vanguards and intellectuals, they have taken into their own hands the task of producing their own knowledge, for their own objectives, with their own parameters of validity and, at times, with alternative sources of legitimization. These are epistemic-political struggles (Sinha-Kerkhoff and Alatas 2010). This crack in the modern order, an apparently fractured and agonizing order, feeds a multiplicity of actors and, with them, a diverse set of knowledges. It is this phenomenon that is analyzed in this volume and for which we propose the notion of epistemic decentralizing as a critical and theoretical suggestion. Before describing the chapters and how the different analyses nourish this theoretical contribution, we first present a tour of the idea of epistemic decentering and decentralizing and then explain how this project builds on, sustains, and relates to others that have focused on the epistemic dimension of social transformations. Finally, we preview the chapters and show the argumentative arc that, with very varied empirical and theoretical contributions, can be observed in this edited volume.

From Centers and Peripheries to (De)centering and (De)centralizing

We will call epistemic centering the process by which some ideas acquire a prominent place within any intellectual space (discipline, field, etc.), eventually constituting a canon that has repercussions on the organization of that same space (Mora et al. 2020; Fasenfest 2022). Centering may or may not be intentional. More often, in fact, it is the consequence of a set of microlevel actions, such as designing courses for a university or launching a book, whose performative character can be perceived only when observed in aggregate. The opposite of centering is decentering. This is an intellectual process (i.e., primarily textual and, to an increasingly lesser extent, audiovisual) by which these central and canonized ideas are challenged (Patel 2021; Restrepo 2016). The challenge can operate through microactions that may eventually constitute a growing trend. The professor who decides to include in her syllabus readings by women philosophers to intersperse their views with those of canonical male philosophers is decentering philosophical thought. A website's list of readings on the intersection of Black feminist thought, culture, and politics can help influence the reading choices of an undergraduate student who is dissatisfied with the content seen in her sociological theory class (Bell 2022). Decentering, then, is the loss of centrality of some mainstream ideas in light of others that relocate, reposition, replace them. What was unique or principal ceases to be so. This is not because the mainstream ideas are unimportant, not because of their intrinsic value, but because of their rearticulation, that is, their relational value (Neubert 2022). To the extent that the relations that sustain ideas change, the ideas themselves change.

The second process, more mundane, more practical, perhaps less interesting for those who take ideas as their focus, is that of epistemic decentralizing. Epistemic decentralizing is the process by which a greater number of actors, of a more diverse nature, with broader interests and more varied resources, acquire the status of full epistemic subjects (Pickering 2005).[1] It can be argued that epistemic decentralizing refers to the distribution of cognitive authority or, in a broader sense, to the distribution of agency itself. If, as Masaka argues, "the acceptance of the epistemic contribution of those who have been historically marked as without any knowledge is necessary if their agency is to be awakened and their liberation realised" (2021, 358), then decentralizing would consist in generating the infrastructures and practices for that distribution of agency to take place.

Epistemic decentralizing has what we might call an indirect link to ideas. When more actors can produce and diffuse knowledge, it cannot be determined in advance that those ideas will contribute to the decentering of thought. For, as Schneider points out, "decentralized technology does not guarantee decentralized outcomes" (2019, 27–28). Rather, it is highly likely that if centering has been successful, epistemic decentralizing will begin with a reinforcement of canonized ideas. Put differently, the first attempts to decentralize may often be, in fact, forms of recentralizing. This is because, in the end, at any specific moment mainstream ideas constitute a parameter of what is thinkable, a limit to what should be discussed within a certain space of knowledge.[2]

However, if epistemic decentralizing continues, it is also to be expected that critical voices of decentering will appear, contributing to the rearticulation of those ideas that were taken as central. As a department hires academics with greater diversity of gender, sexuality, race, or class, it is to be expected that the ideas of that faculty will also gain diversity (Morgan et al. 2022; Mousa 2021). Or, in other words, that new ideas will allow for the decentering of previous ideas. In the same way, as citizen laboratories proliferate and practices known as citizen science are consolidated, nonexperts are contributing to an agenda that is more open to urgent needs and to worldviews that are not always compatible with mainstream institutionalized Western science (Krick and Meriluoto 2022).

First and foremost, epistemic decentralizing is a process of creating, expanding, and consolidating broader infrastructures. Epistemically decentralizing is infrastructuring-in-other-ways to ensure sustained temporal and spatial access of a growing number of cognizing subjects and thinking infrastructures (Kornberger et al. 2019) to the sociotechnical networks of knowledge production. This infrastructure is neither easy to produce nor to maintain. Moreover, it often presents contradictions within itself, given that it is the product of a political project that seeks to decenter through diversifying (Molina Rodriguez-Navas, Lalinde, and Morales 2021). And diversifying inevitably brings conflict.

Epistemic decentralizing is always conflictive and, although it can be thought of as a goal, it could be better thought of as a means, as a condition of production. Without epistemic decentralizing, any effort at epistemic decentering runs several risks simultaneously. The first is to be just a passing fad within mainstream institutions of any space of intellectual production. The second risk is to turn certain already hegemonic actors into nonlegitimate spokespersons for positions, claims, demands, and proposals that do not belong to them. A third risk also appears: decentering can become an empty discourse in which, in the end, nothing is transformed. It can become a discourse that, although full of good intentions, ends up reinforcing the power of the powerful and moving away from the commitment to empower actors to allow them access to the category of full cognitive subjects.[3]

In what follows, I present four scenarios from this vision of centering/decentering on the one hand, and centralizing/decentralizing on the other. As moments in a cycle, one can move toward any of the four scenarios at any time. This implies that an indisputable and permanent progression toward a greater integration of more cognitive subjects or thinking infrastructures in the networks of knowledge production cannot be taken for granted. Rather, it should be thought of as stages within cycles that can follow varied and indeterminate paths, marches, and countermarches.

Defining Epistemic Decentralizing

In the world of ideas, whether in the more radical humanities or in the more positivist natural sciences, the tendency in recent centuries has been toward centralization and centering. The former, intimately related to the action of colonial empires since the fifteenth century, consisted in the elaboration of extensive networks, some of global scope, which allowed the mobilization to their metropolises of objects, nonhuman beings (plants, animals, etc.), and even people from the margins of the empires (Latour 1987; Delbourgo and Dew 2008; Prasad 2014; Stöckelová 2012). Centralization allowed the conformation of the great centers of knowledge accumulation (e.g., libraries, zoos, botanical gardens, archives, etc.) that made cities such as Madrid, London, and Paris imperial capitals and centers of knowledge production (Burke 2000; Jöns 2011). While these sites (universities, academies, publishing houses, etc.) gained relevance and prestige, the centralization of actors and infrastructures led to the centering of the ideas produced there, that is, to their canonization (Mora et al. 2020; Fasenfest 2022). The capacity to abstract and theorize that developed in these places was the fruit of centralization. Once high levels of abstraction were reached, and once the prestige produced by such epistemic dynamics was attained, a process of centering took place that manifested itself in the appearance of ideas with a global scope—the global designs, as Mignolo (2000) calls them. Certain ideas, although produced in specific sites (i.e., as local as any other), nevertheless had the capacity to mobilize through those same networks to explain not only their place of enunciation but the world as a whole (Livingstone 2003). Thus, it can be argued that the centrality of these ideas was due not so much to the lucidity of isolated geniuses but rather to the symbolic-material networks that empires built and producers of knowledge capitalized on almost as efficiently as merchants and bankers. Ideas were woven into the fabric of the global colonial, capitalist, and patriarchal order (Grosfoguel 2022).

In recent decades, there has been a profound revision of the idea of the canon, of the mainstream, of the centrality of some knowledge (Connell 2018; Meghji 2021; Go 2016). As historical, anthropological, and sociological studies showed how much of those central ideas had come from the periphery (Goody 2012), the belief in the universality of knowledge produced in the metropolises (and, for that matter, of any knowledge) was widely criticized. Recognizing the active role of other subjects (peoples, institutions, individuals) in the production of knowledge, as well as the epistemically extractivist nature of colonialism, became central to more recent critical academic approaches (Broncano 2020; Kunce 2022). Ideas and worldviews of indigenous peoples, women and sexual minorities, and other vulnerable groups are gradually being discussed and put into dialogue with institutionalized Western knowledge (Essanhaji and van Reekum 2022; Strasser et al. 2019). Thus begins an attempt to decenter the center (Narayan and Harding 2000) that could be defined as an intellectual hybridization to accommodate a plurality of perspectives and observe how, by incorporating them, more robust knowledge emerges (Harding 1998; Connell 2007).

Undoubtedly, decentering feeds on ideas produced at the margins and channeled toward the center by powerful spokespersons who position themselves politically and epistemically in doing so—sometimes committing epistemic extractivism. Decentering is, in short, a reconfiguration of dominant or hegemonic ideas in light of new ideas that challenge and correct them, giving rise to strong objective knowledge (Harding 2015).

In contrast, epistemic decentralizing happens when the process that began with the collection of objects and the mobilization of people from the periphery to the centers is reversed. Think, for example, of the restitution by museums in the developed world of works of archaeological or historical value to peripheral governments or institutions that claim them (Hicks 2020; Laely 2020). Another example is the creation of scientific infrastructure financed from the centers and installed and operated in marginal regions of the world (Losego and Arvanitis 2008). Yet another is the appearance of journals or editorial initiatives in the periphery that have material and symbolic resources from metropolitan institutions (Albornoz, Okune, and Chan 2020). And there are the mobility grants between countries in the Global South (Sherbondy 2017; Gray and Gills 2016), and the research platforms, such as laboratories, based on feminist values (Kaşdoğan 2020). The list can go on and on. To reverse centralization is to make some of what has been concentrated in the centers flow to the margins, including prestige, economic resources, institutional capacity, or, ultimately, authority.

Reorienting flows does not necessarily imply thinking against the tide or producing counterhegemonic thinking, but simply empowering peripheral actors, infrastructures, and practices in such a way that they can produce knowledge of themselves (and the world; see Keim 2011). There is no guarantee that empowered cognizing subjects will challenge the canonical knowledge of a field or discipline. However, it can be expected that such subjects would acquire sufficient autonomy to think about their agendas, concepts, and methodologies and, with them, address their priority issues (Mousa 2021; Neubert 2022; Patel 2021; Kreimer 2006; Alatas 2001; Rodriguez Medina et al. 2019). Epistemic decentralizing is, in essence, a way of shifting epistemic agency from the centered networks, institutions, and technologies to the peripheries.

More ideas are almost always a challenge to the mainstream. And more actors, infrastructures, and practices are almost always a challenge to the actors, infrastructures, and practices that have (re)produced that mainstream, that are able to benefit from them and that have interests aligned with them (Pickering 2005; Law 2002; Kornberger et al. 2019). For that simple reason, decentralizing is an epistemic and political action. By decentralizing, conflict is created, occasions of confrontation are multiplied, alternative paths for knowledge and its producers are posed, the status quo is altered, and vulnerable subjects are empowered. Needless to say, there is a risk of a clash of forms of knowledge.

Nevertheless, this empowerment is not primarily political in the sense of seeking access to public decision-making. The empowerment that decentralizing allows is epistemic insofar as it increases the capacity of actors to produce their own knowledge. Why restore works of historical value if not to produce knowledge about one's own past? Why build infrastructures based on solidarity and volunteerism if not to produce knowledge different from that which emerges from competing institutions at the heart of neoliberal capitalism? Why create a regional journal if not to influence research agendas and decouple them, even partially, from those imposed by hegemonic countries? Appropriating the means of cognitive production makes sense only if one has the capacity to put these means into operation for purposes that, for lack of a better word, we could call emancipatory.

This is not a minor issue, because the epistemic struggle (which is ultimately political) requires adopting positions that prioritize care, responsibility, and the construction of inter-epistemic links (de Sousa Santos 2014). This means that more important than people's standpoints themselves are the procedures for managing frictions and dissent (Rosenfeld 2019). Thus, epistemic decentralization depends on multiplying infrastructures and practices that articulate peaceful, constructive, and horizontal ways of resolving conflicts. Robust knowledge will not be only that which has a correspondence with reality, but also that which manages to go through the difficult process of producing inter-epistemic consensus (Hessmann Dalaqua 2017).

The idea of epistemic decentralizing that we propose in this volume is at an exploratory stage. It has the capacity to bring together divergent theoretical and disciplinary perspectives, while also allowing us to find commonalities between critical approaches to the current production of knowledge at the global level. It will be seen in the following pages that epistemic decentralizing requires analyzing who and what is at the margins and what are the operations that can be carried out from there. Then it demands looking at the existing and alternative infrastructures, their logics and dynamics, and showing the arrangements that are produced around them. Finally, epistemic decentralizing may give rise to new spaces, often interstitial, weak, little or not at all institutionalized, in which marginal actors *infrastruct* new relations. Given its fragility and relentless enactment, we argue that epistemic decentralizing is always an incompletely realized ambition (Schneider 2019). These relationships may or may not be open challenges to mainstream orderings, but they are always indicative of the agency of these now empowered actors. Such actors take knowledge production into their own hands as an epistemic and political objective (Harding 2016). This journey is reflected in the structure of the present volume, as we will show.

Is Epistemic Decentralizing a New Idea?

Explaining what is novel about the proposal of epistemic decentralization forces us to review a considerable number of efforts, within and outside academia, to address the epistemic problems of marginalization. Given that the history of social movements itself could fit within this historical account, it is best to present analytically and illustratively the initiatives that have shed light on issues similar to those we bring up in this volume. To do so, we present these initiatives in a table in which we cross two variables (table I.1). On the one hand, we have Miranda Fricker's notions of testimonial and hermeneutic injustice. The former "occurs when prejudice causes a hearer to give a deflated level of credibility to a speaker's word" (2007, 1), that is, when a person's agency is reduced due to the questioning of the validity or legitimacy of his or her word. On the other hand, hermeneutical injustice is "the injustice of having some significant area of one's social experience obscured from collective understanding owing to hermeneutical marginalization" (158). In turn, hermeneutical marginalization occurs when there is unequal hermeneutical participation with respect to areas of a certain group's social experience (153). In other words, hermeneutic injustice is visible when a group cannot systematically construct and develop the concepts with which to account for its own life experience and that of the society in which it lives. Put differently, "it is only when there are no pervasive injustices, no systematic roadblocks to the development of knowledge, that responsible agents should be expected to be minimally knowledgeable about themselves, their peers, and the world" (Medina 2013, 128).

On the other hand, we can take two alternative notions of injustice, proposed by Nancy Fraser (2000a, 2000b). Cultural injustice is imbricated

Table I.1 Forms of Injustice and Social Initiatives That Address Them

	Cultural injustice	Socioeconomic injustice
TESTIMONIAL INJUSTICE	#BlackLivesMatter (Gupta and Stoolman 2022) #MeToo (Gurrieri et al 2022) Intersectionality (Hill Collins 2019)	Indignados (Pleyers 2024) Occupy Wall Street (Woodly 2021)
HERMENEUTICAL INJUSTICE	Theory from the South (Comaroff and Comaroff 2015) Feminist standpoint theory (Harding 2016) Provincializing knowledge (Chakrabarty 2000) Decolonizing the canon (Mbembe 2016) Queer academia (Moussawi and Vidal-Ortiz 2020) Decolonial thinking (Mignolo 2000) Postcolonial thought (Go 2016) #CiteBlackWomen (Smith et al 2021)	Epistemic decentralizing Epistemologies of the South (de Sousa Santos 2018)

in the social models of representation, interpretation, and communication and involves cultural domination, lack of recognition, and lack of respect (Fraser 2000a). This is based on the idea that recognition designates a reciprocal relationship between subjects who see themselves as equals, which makes it an essential dimension of the process of development of the sense of self (Fraser 2000b, 109). When a group is devalued by a dominant culture, a phenomenon of misrecognition occurs whereby vulnerable groups internalize a representation of themselves that is imposed and impossible to challenge. A policy of recognition aims to produce self-representations that reaffirm the cultural dimensions of marginalized groups (Fraser 2000b, 109–10). Socioeconomic injustice, on the other hand, is rooted in the economic-political structure of society and includes exploitation, economic inequality, and deprivation (Fraser 2000a). Although, as the author argues, "properly conceived, struggles for recognition can aid the redistribution of power and wealth and can promote interaction and cooperation across gulfs of difference" (Fraser 2000b, 109), analytical differentiation is productive, as we shall see below.

If we combine these forms of injustice, we can produce four scenarios in which we locate initiatives that have struggled against them, in academia or outside of it. Testimonial-cultural injustice has been attacked by collectives that organize testimonies of people, belonging to vulnerable or oppressed sectors, to make them visible in key social instances, such as the media or the courts of justice. These testimonies, rather than constituting alternative theorizing (original conceptualization that refutes or contradicts dominant views), are put into operation in preestablished institutional instances in an attempt to increase the credibility of the members of these groups. Testimonial-socioeconomic injustice has been denounced by initiatives that recognize that the prevailing levels of material inequality (that have increased in the last five decades of rampant neoliberalism) have conditioned the testimonial capacity, of individuals on an equal footing, of an increasingly large sector of global society. The criticisms against the concentration of wealth that have emerged in social movements such as Occupy Wall Street or Indignados, as well as in civil society organizations (Oxfam) or intergovernmental organizations (United Nations) are not limited to pointing out the economic consequences of the enormous impoverishment. They also highlight the loss of the capacity of marginalized actors to organize themselves, participate in decision-making, and contribute to thinking about and implementing forms of wealth redistribution.

A third form of injustice is hermeneutic-cultural, which is perhaps among the best known in academia and to which, in some ways, this volume reacts—without necessarily opposing. Hermeneutic-cultural injustice refers to the inability of some actors, degraded by the mere fact of belonging to denigrated social sectors, to propose with their own words (concepts) a reliable—and politically operative, we might add—description of their social reality. The absence of members of certain social groups in university faculties (women and gender minorities, people with different abilities, racial or ethnic minorities, lower classes, etc.) has been pointed out, to a large extent, as a sign of the need to break with the monolithic dominance of certain groups (e.g., white, Anglo-Saxon, Protestant men). Thus, provincializing knowledge, citing Black women, decolonizing the canon or the curriculum, and theorizing from the South are nothing more than calls to complement (and, in some more radical cases, replace) Eurocentric knowledge with other knowledge produced elsewhere, by other peoples, for tackling other problems. Decolonial thought, feminist perspective theory, and queer academia seek to make visible the intellectual production of socially, politically, and economically marginalized people. All of them, in some way, understand that what is central is to bring the ideas of peripheral groups to the center of the debate, of the disciplines, of the academic, journalistic, artistic, or cultural fields. Important as they have been, they tend to produce two important effects, displacement and reification.

> Questions of recognition are serving less to supplement, complicate and enrich redistributive struggles than to marginalize, eclipse and displace them. I shall call this the problem of displacement. . . . Today's recognition struggles are occurring at a moment of hugely increasing transcultural interaction and communication, when accelerated migration and global media flows are hybridizing and pluralizing cultural forms. Yet the routes such struggles take often serve not to promote respectful interaction within increasingly multicultural contexts, but to drastically simplify and reify group identities. They tend, rather, to encourage separatism, intolerance and chauvinism, patriarchalism and authoritarianism. I shall call this the problem of reification. (Fraser 2000b, 108)

It is in this context that the fourth type of injustice, hermeneutic-socioeconomic, must be read. This form of injustice refers to the systematic inability of certain groups, for reasons of insufficient access to material resources (infrastructures) and consequently inappropriate practices, to produce their own worldviews on the basis of a conceptual (theoretical) development emerging from the periphery. Here it is not so much the testimonial value of socioeconomically oppressed groups that is of interest, but rather their categories of analysis for thinking reality; a reality, it should be said, that not only refers to themselves—as Fricker's proposal of hermeneutic injustice seems to imply in part—but that can also account for the reality of privileged groups and their mechanisms of oppression. In other words, hermeneutic-socioeconomic injustice is what keeps certain groups in society in a noncognizant subject status or, what is the same, what prevents the possibility for subjects to treat each other as peers, in a framework of epistemic humility and solidarity.

In this sense, de Sousa Santos's proposal called "epistemologies of the South" is perhaps the one that has best accounted for the consequences of socioeconomic injustice in their hermeneutic dimension: "The epistemologies of the South concern the production and validation of knowledges anchored in the experiences of resistance of all those social groups that have systematically suffered injustice, oppression, and destruction caused by capitalism, colonialism, and patriarchy. The vast and vastly diversified field of such experiences I designate as the anti-imperial South. It is an epistemological, nongeographical South, composed of many epistemological souths having in common the fact that they are all knowledges born in struggles against capitalism, colonialism, and patriarchy" (2018, 1). Within this epistemology, he proposes a sociology of absence that consists in showing the many practices, knowledges, and agents that exist in nonmetropolitan societies and sociabilities that are actively produced as nonexistent by dominant ways of knowing, especially when confronted with the exclusions proposed by neoliberalism in accepting capitalist, colonial, and patriarchal forms of oppression (de Sousa Santos 2018, 8). His proposal even gives rise to the extreme case in which forms of knowing are directly extinguished within the ecology of knowledges, which he calls epistemicide. There are two central aspects that distinguish the notion of epistemic decentralization from epistemologies of the South.

The first is that epistemic decentralization does not start from the idea that the point of view of the oppressed is always an emancipatory, anticapitalist, anticolonial, and antipatriarchal knowledge. There are plenty of examples of intellectual production in the Global South that reproduce part of the dependence and conceptual subordination that characterizes the peripheral condition. It is not enough, in this sense, to consider that such production is simply the epistemology of the North, given that in many ways it is anchored to conditions of production of the South, in the exact sense given by de Sousa Santos. Put differently, the notion of epistemic decentralization accepts that the inclusion of marginalized sectors in the production of knowledge does not necessarily mean, although it does amplify the chances of, counterhegemonic knowledge.

The second difference is tactical, as Østmo et al. state in chapter 5 of this book. de Sousa Santos understands systems of oppression in a monolithic way and, therefore, overestimates the power of each—and, to some extent and tangentially, problematizes the capacity to confront such systems. Neither patriarchy, nor capitalism, nor colonialism are all-powerful monolithic entities, but sociotechnical systems maintained, with effort, by actors who benefit from their operation and, therefore, from the privileges anchored to such operation. By not reifying these social orders, epistemic decentralization calls for alliances between North and South (and their epistemologies) that make it possible to produce infrastructures and practices that give rise to a diversity and plurality of ideas and knowledge.

Thus, not only does the focus on infrastructure that we propose allow us to distinguish this project from de Sousa Santos's epistemologies of the South, but we also seek to emphasize the relational character of the parties involved in the production of knowledge. What is at the center is, perhaps and in certain places, neoliberal, but it could be otherwise. Epistemic decentralization contributes to other knowledge always having a presence in the debate and intellectual exchange over time and without the need for spokespersons among the powerful actors.

Insofar as all the forms of injustice mentioned above have been denounced and attacked by initiatives such as those included in the table, it is clear that the idea of epistemic decentralization is based on numerous previous experiences and seeks to expand their capacity to attack the main forms of injustice in force. Trying to put back on the table the issue of material asymmetries (in addition to the symbolic ones, almost permanently at the center of attention nowadays), epistemic decentralization connects with traditions of classical critical thought, but highlighting the epistemic (basically, hermeneutic) dimension at stake. Thus, and as the reader will see in the following pages, light will be shed on concepts and experiences that will resonate with the vast literature and casuistry in which epistemic decentralization is embedded.

The Structure of the Book

This volume attempts to answer three main questions. The first is, what are the consequences of the historical process of concentration of knowledge in certain places and subjects (Livingstone 2003) and how can they be reversed? The second asks, what epistemic conditions and sociotechnical arrangements must be established for reversal to take place? The last question is, what do reassemblages that seek to challenge and reconfigure the capitalist, patriarchal, and colonial global order look like? Around these questions, the book is organized in three parts.

The first part explores the cognitive and sociotechnical displacement of thinking produced from the margins. The margin is a concept whose vagueness indicates that there are many forms of concentration (and, therefore, centers) that shape more or less effective, and more or less global, structures of exclusion. Every process of concentration, whether material or symbolic, ends up forming a center that capitalizes on some kind of resource and a periphery or margin that is relegated and, to a certain extent, exploited. Chapters 1 through 4, by Alcoff, Rodriguez Medina, Kleinman, and Traweek, explore these marginal positions. These chapters map the consequences of the historical concentration of knowledge, reviewing extractive epistemologies, the social production of ignorance, and the voices that, while situated in the center, consciously move to marginal positions. Taken together, the chapters review central debates in current science and technology studies and in epistemology and sociology of knowledge that have focused on epistemic asymmetries caused by enduring structures of oppression, such as patriarchy, colonial order, or cognitive capitalism that permeate social life. They contribute to the conceptual framework (in particular Alcoff and Rodriguez Medina) but also show connections to widely discussed literature in the field (e.g., sociology of ignorance) and advance methodological recommendations, for example by claiming the value of certain practices (gossip or joke) to make marginal positions observable within academic fields.

Alcoff reflects on the effect of colonialism and imperialism on knowledge practices, arguing that extracting resources, material or intellectual, always involves an epistemic component (chapter 1). In a propositional manner, the author suggests four corrective epistemic norms that seek to replace extractivist epistemologies with collaborative forms of knowledge production. They are (1) acknowledging the incompleteness of all knowledge, (2) developing approaches that recognize plural epistemologies and seek productive relationships of inter-epistemology, (3) practicing relational epistemic humility, and (4) regularizing the assessment of epistemic relationships in projects of knowing. The relevance of Alcoff's argument for understanding epistemic decentralization is twofold. On the one hand, it brings to the forefront the ultimate reason why the process of centering of ideas takes place: they are valuable, in the sense of commodifiable, within contemporary extractive capitalism. On the other hand, by recognizing that knowledge production practices need to be sustained by certain epistemological ideas, Alcoff is showing a connection between decentering and decentralization: justification. Thus, for epistemic decentralization to occur, it follows from this chapter that a profound revision of the postulates behind the current extractivist epistemology that characterizes the metropolitan academy is needed.

In a similar vein, Rodriguez Medina argues that, in order to understand the relationship between centers and peripheries of knowledge production, it is necessary to combine two variables: the degree of decentering of ideas and the degree of decentralizing of epistemic infrastructures and practices (chapter 2). This proposal allows him to construct four types of links, which he calls (1) epistemic indifference, (2) epistemic co-optation, (3) epistemic extractivism, and (4) epistemic mutualism. In some sense, it can be argued that the corrective norms pointed out by Alcoff are the mechanisms to reach the phase of epistemic mutualism that is inferred as overcoming in Rodriguez Medina's chapter on epistemic decentralizing. The stage of epistemic mutualism appears, in the chapter, not as an inevitable destiny but as a possible scenario to which academic fields or disciplines may or may not be heading. Moreover, the cases show that epistemic decentralization is never a finished process, but one that is constantly being updated. In this sense, knowledge production is always in some process by which actors are empowered or disempowered as epistemic subjects, which allows us to think of the theoretical concept suggested as an evaluative framework of concrete epistemic initiatives.

Kleinman, for his part, illustrates the effects on ignorance of the decentralization of actors and practices and the decentering of agendas (chapter 3). In the case of AIDS treatment research, under pressure from patients/activists, scientists considered new models of clinical trials and allowed new lines of research to emerge. Thus, the ignorance that can result from unrealized science was eliminated, and more comprehensive and robust knowledge was produced. In contrast, the knowledge production of beekeepers went unnoticed, unrecognized, or ignored by mainstream scientists, regulators, and corporate stakeholders. As a consequence, a whole area of decentered research concerns and questions remained unknown due, in no small part, to the degree of centralization imposed by conventional science practices and actors. The chapter's main contribution to the understanding of epistemic decentralization is that more actors and more perspectives do not always translate into more knowledge. Kleinman's cases teach that the acceptance and validation of knowledge produced by peripheral actors rests, in no small measure, with powerful actors in the field working as gatekeepers to new understandings. Without the epistemic humility that Alcoff and Rodriguez Medina bring to light in their chapters, it is practically impossible to assume that the mechanisms through which knowledge from the margins is accepted in the center will be constituted, respected, and enhanced.

The first section culminates with Traweek's personal account, which has profound institutional implications for understanding epistemic decentralizing at a micro scale (chapter 4). An expert in the construction of scientific authority, she argues that we must study practices at the edge of epistemic authority to understand how that authority is constructed, maintained, and sometimes subverted. Hence the relevance of stories, gossip, whisper cultures, and even jokes, which are not made or circulated in the same way as institutionalized knowledge. It is emphasized in the chapter that the shift to the margins is not only a product of epistemic asymmetries in operation, but also, at times, the conscious decision to seek places of enunciation more apt for critical thinking of those asymmetrical structures. With Traweek's chapter, epistemic decentralization becomes not only a process of opening up to marginal ideas and actors, but a strategy of some powerful actors to disrupt mainstream ideas. The importance of this approach cannot be underestimated. On the one hand, marginalization can be voluntary (against that imposed by an unequal structure of knowledge production) and, even so, serve the same ends: de-canonizing, decentering. On the other hand, when betting on this strategy, central actors of the field or discipline must opt for subtle, fragile, deinstitutionalized, almost ethereal actions, with which the invisible might become visible.

The second part of the book analyzes new infrastructures that make it possible to challenge asymmetries of knowledge and power. The chapters in this section produce double analyses. On the one hand, they show how current structures in certain fields, such as academia, political decision-making, or agroecological innovation, tend toward cognitive, political, and economic centralization respectively. In pointing this out, they take the approaches of the first section a step further by showing that marginal positions are reproduced through complex technologies that are very difficult to disassemble, or to reassemble. Moreover, the chapters show how the epistemic norms behind knowledge production practices, analyzed in the first chapters, are imbricated in the technologies and infrastructures that make them viable. Thus, the political nature of the processes of centralization and decentralization becomes evident. In the face of forms of domination by powerful actors (be it the Norwegian state, large Brazilian agricultural producers, or multinational academic publishers), the work of infrastructuring otherwise becomes an act of ethico-political resistance. At the same time, the chapters illustrate that alternative infrastructuring has high costs, touches on entrenched interests in the sociopolitical and economic order, and is often precarious in nature, which also complicates the possibility of making them paradigmatic cases in other contexts.

Østmo et al. focus on Sápmi narratives to show the contradictions between ancestral indigenous practices and contemporary political arrangements in Norway that do not capture (and, in fact, put at risk) ancestral forms of survival and their cultural manifestations (chapter 5). Østmo et al.'s analysis of epistemic decentralization may seem contradictory. While acknowledging that centering, decentering, centralizing, and decentralizing are tangled together (something science, technology, and society [STS] has taught us as we move freely in sociotechnical networks), they also point out that knowledge practices have performativity—that is, they produce an effect beyond the content of that knowledge. It is infrastructural practices that allow the researcher to see hidden agendas, a task that includes looking at the politics of language, meeting dynamics, archives, reports, and other mundane forms of knowledge materialization. But in doing so, the chapter leaves two fundamental corollaries for continuing to decentralize knowledge: (1) there are no monolithic centers (nor mainstream, nor canons) and, in their inconsistencies, there is room for alliances; and (2) decentralizing is a tactical matter, that is, there are no definitive rules but opportunities. The heterogeneous and fragile, but also powerful and transformative, nature of decentralization processes is evident in this chapter.

Okune et al. provide numerous examples of ways of dealing with the processes of infrastructural concentration that affect the current academy (chapter 6). The academy seems tied to practices and technologies that favor exclusion, such as article publication charges from large publishers or forced internationalization at the university level. However, initiatives such as IstanbuLab in Turkey and Kaleidos in Ecuador show that other forms of infrastructure are both possible and desirable. Okune and her colleagues demonstrate that, when epistemic decentralization begins to operate—what the authors call remooring—the fields or disciplines tend to retrenchment, that is, to return to a certain previous status quo, to reestablish hierarchies, to reconstruct canons. Not only does this show that decentralizing and decentering do not always go hand in hand, as Rodriguez Medina shows in his classification in chapter 2, but that many times decentralizing and decentering will be judged in the light of current mainstream ideas, with their metrics and assessment mechanisms. Moving away from canonized ideas, then, is a tedious task in which there is a back-and-forth between the new and the old and which, always, involves infrastructuring in other modes.

Levidow, in chapter 7, shifts the focus and shows that decentering and decentralizing of knowledge are also economic and political issues that touch food and nutrition, as well as other areas. Looking at solidarity economy initiatives in Brazil, Levidow describes short supply chains that seek to connect more closely producers and consumers, whose purchases come to play not only an economic but also a political role, given that they support cooperative work organization and environmentally sustainable practices. Thus, it is clear that epistemic decentralizing must transcend the academy and contribute to rethinking how to reorganize work and consumption, even within the limited liminal spaces of global neoliberal capitalism. Levidow's chapter brings clarity and precision to a key aspect of understanding epistemic decentralization. Equivalent to the idea of epistemic humility that Alcoff and Rodriguez Medina claim for the academy, Levidow concludes that solidarity within the extended networks of agrifood producers and consumers is central to epistemic decentralization. By seeking to decenter agrifood knowledge by the mainstream economy and to decentralize the infrastructures that make that knowledge possible, *circuitos curtos* of producers and consumers construct affective sociocultural meanings that reconnect people from different contexts and localities. Without these affective networks, it seems impossible that the meanings produced by the new actors included in the networks can move and, ultimately, challenge ideas accepted as truths within certain spheres—such as agribusiness. And, as expressed by Okune et al., Levidow points out that the opportunities opened up by recent crises, especially the COVID-19 pandemic and climate change, are not an invitation to return to a status quo that is seen as part of the problem. As the case shows, inasmuch as decentralization depends on solidaristic economic forms, this process could be considered a response to the need to reorder and reorganize social life, to a large extent to make room for other forms of knowledge.

In the third and final section, it is shown that when at work, infrastructures and practices supporting voices from the margins end up decentralizing knowledge by producing new sites of knowledge production. These spaces, originally interstitial within spaces governed by hegemonic practices, institutions, and actors, are proof that alternatives to current sociotechnical orders, whether at the level of global powers or mainstream medicine, as Østmo et al. show in chapter 5, are always unstable and fragile. The chapters in part III show that, from infrastructures and practices, there is a boundary work that delimits alternative spaces, in line with Okune et al.'s findings. These new spaces, sometimes with preexisting logics (as in the intergovernmental conferences studied by Keim and Sitas, chapter 10) and sometimes with new dynamics (such as the palliative care centers in Taiwan analyzed by Kuo, chapter 8), have the difficult task of articulating with other spaces, either challenging them (mainstream) or producing strategic alliances (peripheries). Likewise, spatialization allows us to think of scales for the process of epistemic decentralizing. While in some cases this process may involve individuals or organizations (as shown in chapter 8), in others it occurs at the national (as illustrated in chapter 9, by Cancino et al.) or global scale (as pointed out by Keim and Sitas focusing on the BRICS countries). The mechanisms inherent at each level, however, are different and invite reflection on the growing political will that is necessary to convert a process that has been basically intended for understanding academic fields' and disciplines' organization to the level of international geopolitics.

Kuo shows that such spaces for the production of counterhegemony—or, at least, of ideas that challenge the mainstream—can be institutionalized. Based on fieldwork in a Taiwanese medical center, Kuo presents palliative care as a decentralized profession. Palliative care wards emerge both as therapeutic and as knowledge spaces that nurture the profession. This does not mean that the biomedical scheme should be overridden, but rather that it should be understood as part of the total effort to achieve a favorable end of life. The chapter opens the door to think of epistemic decentralization not only as a way of making a field or a discipline more plural, but also as a way of practicing a profession in a different way. As in chapter 7 (Levidow), the disciplines are transcended to enter the world of practitioners who seek to reconfigure a space based on practices that, in principle, have no validity in the core of the medical field. What is remarkable in the chapter is that, even in the difference, the epistemic perspectives do seem to find a common ground: the observation of the care of patients in the terminal phases of their illnesses. Thus, the chapter sheds light on one way of evaluating the reorganization of a field, as Okune et al. also showed: by paying attention to the type of outcome that the new practices give rise to. Far from being an incompatible and incommensurable alternative, the decentralized knowledge produced in palliative care wards is inserted into the biomedical sciences, that is to say, they maintain a dialogue, insofar as both are perceived as complementary to guarantee a dignified end of human life.

Cancino et al. go a step further and take the issue of epistemic decentralizing to the national level, with the case of the current political reforms in Chile that are intended to decentralize its science and technology system (chapter 9). The authors have found that certain relevant issues in these territories, such as salmon farming and forest fires, show how local demands generate the decentering of scientific agendas through processes of decentralizing of actors, capacities, and decision-making. In other words, socioenvironmental conflicts and/or sociotechnical controversies decentralize local knowledge, activate citizens and governments, and contribute to the production of social and political changes that generate decentered agendas linking local priorities with global issues. The findings in this chapter present a challenge to the conceptualization presented in this introduction and in conceptual chapters in the first section of the book. Here it is the decentering of ideas (attention to a local agenda of problems) that leads to the decentralization of knowledge (through greater stakeholder participation). How is it possible that highly concentrated forms of knowledge production even allow the emergence of a decentered agenda? The answer is in the chapter itself and is key to understanding the transnational nature of any process of epistemic decentralization. Cancino et al. find that the transnational character of local problems is fundamental to understanding the will to decentralize knowledge production. Differently put, the local agenda of salmon farming and forest fire experts has an inescapable international dimension that prevents the understanding of these issues from an exclusively local perspective. Thus, the chapter would suggest that the less local a problem is, the more easily decentralizable is the network of actors, practices, and infrastructures that can address it.

The section ends by taking the question of epistemic decentralizing to the highest level, that of the geopolitical rearticulation brought about by the appearance of emerging powers, the BRICS (Brazil, Russia, India, China, and South Africa). Keim and Sitas argue that beyond economic and international relations policy, the BRICS countries have set out to decentralize science and the global academic scene and have created particular institutions that connect actors from previously nonhegemonic countries (chapter 10). In a twist that seems paradoxical but is not, the authors show a decentralizing strategy led by states that have tended to centralize their decision-making processes in relation to the formation and extension of the BRICS alliance. However, the key is that through this centralized decision-making, states seek to empower subnational actors (universities, professional associations, national academies, etc.) in the process of beginning to construct a vision of themselves beyond that which has been imposed on them from the centers of knowledge production. The BRICS countries know that they are acting (and thinking) from the straitjacket of Eurocentric knowledge and that this limits their possibilities of understanding themselves, their global alliance, and their possible actions to reconfigure the world order. Hence, epistemic decentralizing would aim, at the end of the road, to enable the construction and dissemination of their own representations. By showing and also criticizing political centralization as a condition for the existence of decentralized epistemic infrastructures and practices, the chapter raises urgent questions about the best strategies for producing non-Eurocentric knowledge—questions whose resolution is beyond the scope of this volume.

The three sections are articulated to show possible paths and present any potential obstacles to them. They combine micro and macro perspectives. They come from empirical work, philosophical reflections, and historical analysis. They answer key questions but also raise deeper ones. In short, they begin to fill a void that our three trigger questions sought to bring to the fore: how knowledge has been concentrated and centralized, what arrangements allow both processes to be reversed, and what such rearrangements do or could look like. It is no small matter that, in many cases, the chapters have been written by author collectives that, in themselves, are an example of epistemic decentralization, combining generations, nationalities, disciplinary backgrounds, and, consequently, standpoints. Similarly, the fact that this volume is available in open access, thanks to the generous support of UCLA, is also indicative of the commitment of the editors and chapter authors to avoid forms of centralization, which, in the end, also leads to a certain (albeit unintended) centering of the debate. Instead, the multiple, rich, and sometimes contradictory answers that each chapter provides to the questions that triggered this collective reflection open up the debate and invite further exploration.

Notes

1   One of the few works in which decentering and decentralizing are used for the study of the production and circulation of knowledge is Mora et al. (2020). There, unlike in the present chapter, a slightly more limited use is made. "When we talk about decentering and decentralizing as two related ideas, we emphasize the need to move the conversation away from historically dominant groups (decenter) and geographical locales (decentralize)" (Mora et al. 2020, 313). This chapter recognizes practical dimensions of both processes, such as the use of a lingua franca, the marginalization of epistemologies from the Global South in educational contexts, the conceptualization of what is valid data, and how research hierarchies are established. Infrastructures, however, are neglected.
2   It could be argued that a key characteristic of centers of knowledge production is diversity and plurality in terms of the ideas they radiate or, in

other words, their capacity to produce mainstream (plural) ideas. In that sense, advanced epistemic decentralization is likely to produce not only counterhegemonic ideas (Keim 2011) but rather a multiplicity of influential ideas, some of which may present challenges to previous canons and others of which may reinforce or extend them.

3   The consequences of this disempowerment cannot be underestimated. For example, a recent study has pointed out that "Africa's emigration crisis is traceable, inter alia, to the epistemic imbalance in the very structure of modernity. This imbalance results from the stifling of Africa's epistemic resources under Western epistemic hegemony. Epistemic coloniality, of course interacting with some material factors, creates a sufficient condition for emigration" (Ude 2022, 3).

References

Alatas, Syed Farid. 2001. "The Study of the Social Sciences in Developing Societies: Towards an Adequate Conceptualization of Relevance." *Current Sociology* 49 (2): 1–19.

Albornoz, Denisse, Angela Okune, and Leslie Chan. 2020. "Can Open Scholarly Practices Redress Epistemic Injustice?" In *Reassembling Scholarly Communications: Histories, Infrastructures, and Global Politics of Open Access*, edited by Martin Paul Eve and Jonathan Gray, 65–79. Cambridge, MA: MIT Press.

Bell, Carol V. 2022. "5 Books at the Intersection of Black Feminist Thought, Culture, and Politics." NPR, March 21. https://www.npr.org/2022/03/21/1087494475/5-books-at-the-intersection-of-black-feminist-thought-culture-and-politics.

Blair, Ann M. 2011. *Too Much to Know: Managing Scholarly Information before the Modern Age*. New Haven, CT: Yale University Press.

Broncano, Fernando. 2020. *Conocimiento Expropiado: Epistemología Política en una Democracia Radical*. Madrid: Akal.

Burke, Peter. 2000. *Social History of Knowledge: From Gutenberg to Diderot*. Cambridge: Polity.

Chakrabarty, Dipesh. 2000. *Provincializing Europe: Postcolonial Thought and Historical Difference*. Princeton, NJ: Princeton University Press.

Comaroff, Jean, and John Comaroff. 2015. *Theory from the South: Or, How Euro-America Is Evolving toward Africa*. London: Routledge.

Connell, Raewyn. 2007. *Southern Theory: Social Science and the Global Dynamics of Knowledge*. Cambridge: Polity.

Connell, Raewyn. 2018. "Decolonizing Sociology." *Contemporary Sociology* 47 (4): 399–407.

Delbourgo, James, and Nicholas Dew, eds. 2008. *Science and Empire in the Atlantic World*. New York: Routledge.

de Sousa Santos, Boaventura. 2014. *Epistemologies of the South: Justice against Epistemicide*. New York: Routledge.

de Sousa Santos, Boaventura. 2018. *The End of the Cognitive Empire: The Coming of Age of the Epistemologies of the South*. Durham, NC: Duke University Press.

Essanhaji, Zakia, and Roger van Reekum. 2022. "Following Diversity through the University: On Knowing and Embodying a Problem." *Sociological Review* 70 (5): 882–900.

Fasenfest, David. 2022. "Spiking the Sociological Canon." *Critical Sociology* 48 (4–5): 549–52.

Fraser, Nancy. 2000a. "¿De la redistribución al reconocimiento? Dilemas de la justicia en la era 'postsocialista.'" *New Left Review*, no. 1 (January–February): 126–55.

Fraser, Nancy. 2000b. "Rethinking Recognition." *New Left Review*, no. 3 (May–June): 107–20.

Frickel, Scott, Sahra Gibbon, Jeff Howard, Joanna Kempner, Gwen Ottinger, and David J. Hess. 2010. "Undone Science: Charting Social Movement and Civil Society Challenges to Research Agenda Settings." *Science, Technology, and Human Values* 35 (4): 444–73.

Fricker, Miranda. 2007. *Epistemic Injustice: Power and the Ethics of Knowing*. New York: Oxford University Press.

Go, Julian. 2016. *Postcolonial Thought and Social Theory*. New York: Oxford University Press.

Goody, Jack. 2012. *The Theft of History*. Cambridge: Cambridge University Press.

Gray, Kevin, and Barry K. Gills. 2016. "South–South Cooperation and the Rise of the Global South." *Third World Quarterly* 37 (4): 557–74.

Grosfoguel, Ramón. 2022. *De la sociología de la descolonización al nuevo antimperialismo decolonial*. Mexico City: Akal.

Gupta, Akhil, and Jessie Stoolman. 2022. "Decolonizing US Anthropology." *American Anthropologist* 124: 778–99.

Gurrieri, Lauren, Andrea Prothero, Shona Bettany, Susan Dobscha, Jenna Drenten, Shelagh Ferguson, Stacey Finkelstein, Laura McVey, Nacima Ourahmoune, Laurel Steinfield, and Linda Tuncay Zayer. 2022. "Feminist Academic Organizations: Challenging Sexism through Collective Mobilizing across Research, Support, and Advocacy." *Gender, Work and Organization*, October 12, 1–22. https://doi.org/10.1111/gwao.12912.

Harding, Sandra. 1998. *Is Science Multicultural? Postcolonialisms, Feminisms, and Epistemologies*. Bloomington: Indiana University Press.

Harding, Sandra. 2015. *Objectivity and Diversity: Another Logic of Scientific Research*. Chicago: University of Chicago Press.

Harding, Sandra. 2016. *Whose Science? Whose Knowledge? Thinking from Women's Lives*. Ithaca, NY: Cornell University Press.

Hess, David J. 2009. "The Potentials and Limitations of Civil Society Research: Getting Undone Science Done." *Sociological Inquiry* 79 (3): 306–27.

Hessmann Dalaqua, Gustavo. 2017. "Democracy and Truth: A Contingent Defense of Epistemic Democracy." *Critical Review* 29 (1): 49–71.

Hicks, Dan. 2020. *The Brutish Museums: The Benin Bronzes, Colonial Violence and Cultural Restitution.* London: Pluto.

Hill Collins, Patricia. 2019. *Intersectionality as Critical Social Theory.* Durham, NC: Duke University Press.

Jöns, Heike. 2011. "Centre of Calculation." In *The SAGE Handbook of Geographical Knowledge,* edited by John Agnew and David N. Livingstone, 158–70. London: Sage.

Kaşdoğan, Duygu. 2020. "Feminist Laboratuvarda Bilim, Emek ve Politika." *Praksis* 52: 157–80.

Keim, Wiebke. 2011. "Counterhegemonic Currents and Internationalization of Sociology: Theoretical Reflections and an Empirical Example." *International Sociology* 26 (1): 123–45.

Kornberger, Martin, Geoffrey C. Bowker, Julia Elyachar, Andrea Mennicken, Peter Miller, Joanne Randa Nucho, and Neil Pollock. 2019. *Thinking Infrastructures: Research in the Sociology of Organizations.* Bingley, UK: Emerald.

Kreimer, Pablo. 2006 "¿Dependientes o integrados? La ciencia latinoamericana y la nueva división internacional del trabajo." *Nómadas* 24: 199–212.

Krick, Eva, and Taina Meriluoto. 2022. "The Advent of the Citizen Expert: Democratising or Pushing the Boundaries of Expertise?" *Current Sociology* 70 (7): 967–73.

Kunce, Aleksandra. 2022. "Why Should We Cultivate 'the Difference' in Everyday Practices of the University?" *Higher Education Quarterly* 76: 671–82.

Laely, Thomas. 2020. "Restitution and Beyond in Contemporary Museum Work: Reimagining a Paradigm of Knowledge Production and Partnership." *Contemporary Journal of African Studies* 7 (1): 17–37.

Latour, Bruno. 1987. *Science in Action: How to Follow Scientists and Engineers through Society.* Cambridge, MA: Harvard University Press.

Law, John. 1992. "Notes on the Theory of the Actor-Network: Ordering, Strategy, and Heterogeneity." *Systems Practice* 5 (4): 379–93.

Law, John. 2002. *Aircraft Stories: Decentering the Object in Technoscience.* Durham, NC: Duke University Press.

Law, John, and John Urry. 2004. "Enacting the Social." *Economy and Society* 33 (3): 390–410.

Livingstone, David N. 2003. *Putting Science in Its Place: Geographies of Scientific Knowledge.* Chicago: University of Chicago Press.

Lobo, Michelle. 2022. "Breathing Spaces of Fearlessness and Generosity in the Anglophone/Western University." *Geographical Research* 60 (1): 126–37.

Losego, Philippe, and Rigas Arvanitis. 2008. "La science dans les pays non-hégémoniques." *Revue d'Anthropologie des Connaissances* 2 (3): 343–50.

Masaka, Dennis. 2021. "Knowledge, Power, and the Search for Epistemic Liberation in Africa." *Social Epistemology* 35 (3): 258–69.

Mbembe, Aquille Joseph. 2016. "Decolonizing the University: New Directions." *Arts and Humanities in Higher Education* 15 (1): 29–45.

Medina, José. 2013. *The Epistemology of Resistance: Gender and Racial Oppression, Epistemic Injustice, and Resistant Imaginations.* New York: Oxford University Press.

Meghji, Ali. 2021. *Decolonizing Sociology: An Introduction.* Cambridge: Polity.

Meghji, Ali, and Sophie Marie Niang. 2022. "Between Post-racial Ideology and Provincial Universalisms: Critical Race Theory, Decolonial Thought and COVID-19 in Britain." *Sociology* 56 (1): 131–47.

Mignolo, Walter. 2000. *Local Histories/Global Designs: Coloniality, Subaltern Knowledges, and Border Thinking.* Princeton, NJ: Princeton University Press.

Molina Rodríguez-Navas, Pedro, Johamna Muñoz Lalinde, and Narcisa Medranda Morales. 2021. "Interactive Maps for the Production of Knowledge and the Promotion of Participation from the Perspective of Communication, Journalism, and Digital Humanities." *ISPRS International Journal of Geo-Information* 10 (11): 722. https://doi.org/10.3390/ijgi10110722.

Mora, Raúl Alberto, Gerald Campano, Ebony Elizabeth Thomas, Amy Stornaiuolo, Bethany Monea, Ankhi Thakurta, and James Joshua Coleman. 2020. "Decentering and Decentralizing Literacy Studies: An Urgent Call for Our Field." *Research in the Teaching of English* 54 (4): 313–17.

Morgan, Allison C., Nicholas LaBerge, Daniel B. Larremore, Mirta Galesic, Jennie E. Brand, and Aaron Clauset. 2022. "Socioeconomic Roots of Academic Faculty." *Nature Human Behaviour* 6 (12): 1625–33.

Mousa, Mohamed. 2021. "Academia Is Racist: Barriers Women Faculty Face in Academic Public Contexts." *Higher Education Quarterly* 76 (4): 741–58.

Moussawi, Ghassan, and Salvador Vidal-Ortiz. 2020. "A Queer Sociology: On Power, Race, and Decentering Whiteness." *Sociological Forum* 35 (4): 1272–89.

Narayan, Uma, and Sandra Harding. 2000. *Decentering the Center: Philosophy for a Multicultural, Postcolonial, and Feminist World.* Bloomington: Indiana University Press.

Neubert, Dieter. 2022. "Do Western Sociological Concepts Apply Globally? Toward a Global Sociology." *Sociology* 56 (5): 930–45.

Nieto Olarte, Mauricio. 2019. *Una historia de la verdad en Occidente: Ciencia, arte, religión y política en la conformación de la cosmología moderna.* Bogotá: Fondo de Cultura Económica and Universidad de los Andes.

Patel, Sujata. 2021. "Sociology's Encounter with the Decolonial: The Problematique of Indigenous vs That of Coloniality, Extraversion and Colonial Modernity." *Current Sociology* 69 (3): 372–88.

Pickering, Andrew. 2005. "Decentering Sociology: Synthetic Dyes and Social Theory." *Perspectives on Science* 13 (3): 352–405.

Pleyers, Geoffrey. 2024. "For a Global Sociology of Social Movements: Beyond Methodological Globalism and Extractivism." *Globalizations* 21 (1): 183–95. https://doi.org/10.1080/14747731.2023.2173866.

Prasad, Amit. 2014. *Imperial Technoscience: Transnational Histories of MRI in the United States, Britain, and India.* Cambridge MA: MIT Press.

Restrepo, Eduardo. 2016. "Descentrando a Europa: Aportes de la teoría postcolonial y el giro decolonial al conocimiento situado / Decentralizing Europe: Contributions of Postcolonial Theory and the Decolonial Shift to Situated Knowledge." *Revista Latina de Sociología* 6 (1): 60–71.

Rodriguez Medina, Leandro, Hugo Pablo Ferpozzi, Juan Agustín Layna, Emiliano Martin Valdez, and Pablo Kreimer. 2019. "International Ties at Peripheral Sites: Co-producing Social Processes and Scientific Knowledge in Latin America." *Science as Culture* 28 (4): 562–88.

Rosenfeld, Sophia. 2019. *Democracy and Truth: A Short History.* Philadelphia: University of Pennsylvania Press.

Schneider, Nathan. 2019. "Decentralization: An Incomplete Ambition." *Journal of Cultural Economy* 12 (4): 265–85.

Sherbondy, Kelsey. 2017. "African Diaspora Scholar Mobility Programs: Looking toward Models for South-South Cooperation." *Journal of Comparative and International Higher Education* 9 (Winter): 46–47.

Sinha-Kerkhoff, Kathinka, and Syed Farid Alatas, eds. 2010. *Academic Dependency in the Social Sciences: Structural Reality and Intellectual Challenges.* New Delhi: Manohar.

Smith, Christen A., Erica L. Williams, Imani A. Wadud, and Whitney N. L. Pirtle. 2021. "Cite Black Women: A Critical Praxis (a Statement)." *Feminist Anthropology* 2 (1): 10–17.

Stöckelová, Tereza. 2012. "Immutable Mobiles Derailed: STS, Geopolitics, and Research Assessment." *Science, Technology, and Human Values* 37 (2): 286–311.

Strasser, Bruno J., Jérôme Baudry, Dana Mahr, Gabriela Sanchez, and Elise Tancoigne. 2019. "'Citizen Science'? Rethinking Science and Public Participation." *Science and Technology Studies* 32 (2): 52–76.

Ude, Donald Mark C. 2022. "Coloniality, Epistemic Imbalance, and Africa's Emigration Crisis." *Theory, Culture and Society* 39 (6): 3–19.

Woodly, Deva R. 2021. *Reckoning: Black Lives Matter and the Democratic Necessity of Social Movements.* New York: Oxford University Press.

| Thinking from the Margins

Extractivist Epistemologies

This chapter develops the concept of extractivist epistemology as a way to reveal and reflect upon the effect of colonialism and imperialism on practices of knowing. Projects that aim to extract resources of various types, whether material or intellectual, always involve a knowledge component. This knowledge can be pursued by different sorts of practices, but I argue here that the colonial context of extractivism, in all its permutations, has generated certain types of practices and related ideas about epistemic justification that need to be rethought. The epistemic approaches this chapter is concerned with are those which are mainly pursued by agents from the Global North, but they have influenced normative epistemic ideals and epistemic presuppositions more widely.

This chapter is organized into four sections. First, I give a brief overview of extractivism as both practice and idea. Second, I give an initial explanation of the knowledge practices and ideas that grew out of extractivism. Third, I discuss two case studies that reveal the concrete and specific ways in which extractivism negatively impacts knowledge practices. And fourth, I conclude with four corrective norms to counteract extractivist epistemologies. Situated as we still are within a world formed by colonialism, in terms of ideas as well as material relations, reforming epistemic practices needs to follow a nonideal approach. This means that alternative norms are understood as corrective and ameliorative in relation to current injustices rather than timeless and universal (Khader 2019; Mills 2005).

Capitalism continues to pursue projects that extract material resources from the formerly colonized areas of the world. These include mineral and plant resources as well as timber, fossil fuels, and animal products that are then monetized and exchanged. This form of capital accumulation was driven by the needs of colonial and imperial powers since the beginning of the Conquest. Then, as now, it was cloaked by terms like *progress, development,* and benevolent *stewardship.*

Extractivist projects of resource accumulation transform not only economies but also political and legal institutions and the organization of labor, resulting in a transformation of social identities and communal relations (Bebbington 2010; Escobar and Pardo 2007; Harvey 2003, 2004; Petras 2012; World Bank 2005; Veltmeyer and Petras 2014). As Robinson explains, the result of integrating local economies into global capitalism is that "existing social relations are disarticulated and replaced by new sets of relations shaped by the commercial, productive and cultural processes of global society" (2008, 57). Extraction often results in territorial dispossession but also a degradation of the environment, including at the subsoil level, that leads to loss of livelihoods (Veltmeyer and Petras 2014). Because of the dangers posed to the climate by these practices, extractivist capitalism is now engendering widespread critical analysis and collective resistance beyond the immediately affected communities.

To be sure, extractivism is a broad concept applicable to many sorts of projects, including those generated by formerly colonized nations (Riofrancos 2020). Even in regard to transnational and neocolonial economic activity, the term can be misleading: the era of primitive accumulation, as it was practiced in the early days of the Conquest, is no longer the main form extraction takes. Forms of "biopiracy" and "bioprospecting," for example, aim not only to extract but to patent plant material and control its production and distribution, as is discussed in one of the case studies below. What is taken is not simply a product but a process of production. Also, there are new expansions of extractivism in the arenas of debt, real estate, and social media.

Another new aspect of extractivism involves the fact that it is not generally part of a political takeover of a nation-state (though coups continue to be perpetrated in covert operations), but that it nonetheless leads to a reorganization of entire regional economies. As Robinson notes, "Each cycle of integration into world capitalism is also associated with an extension of capitalist institutions and production relations in the region" (2008, 51). New forms of economic colonialism work to integrate both regions and local communities into transnational capital markets in ways that benefit and enrich the Global North but are also administered by local elites to their advantage.

After achieving independence from colonial powers, many societies developed agro-export-based economies in which local goods of various sorts were monetized and traded. But in the twentieth century, transnational markets began to determine what types of agricultural products were grown. Import substitution industrialization largely replaced subsistence and variegated farming with monocrops such as palm oil and soy. Today import substitution has itself been overtaken by a more general export-led development that "favors new circuits of production and circulation linked to the global economy," displacing peasant agricultural communities and causing social unrest (Robinson 2008, 54).

Despite variation, what links most types of extractivist projects today is an orientation that treats both land and peoples primarily as resources. Seen primarily as resources, land, timber, bio-rich plants, labor, and communities are subject to external reorganization, without participatory decision-making, guided only by the desire for more profit.

It is important to emphasize that transnational corporate entities enrich not only foreign nationals but also domestic elites. Mining projects can also bring significant employment as well as large payments, which may be used by progressive governments for public goods. Poorer nations that have rich mineral deposits, such as Ecuador, Colombia, and Bolivia, are thus caught up in a competition to attract transnational corporations (Sankey 2014; Riofrancos 2020). However, the principal leverage governments can offer in this competition for foreign investment is control: "allowing for unilateral expropriation of private property," with minimal restrictions or state participation (Sankey 2014, 123). State-run mining and coal companies, along with their associated unions, have largely been liquidated, as Sankey reports, "under pressure from the IMF."

Extractivist projects embedded in a neocolonial global economy are continuing many of the ideas and practices that began with the Conquest of the Americas. Still today, there is little egalitarian collaboration or shared decision-making, and few states mandate cooperation between interested parties or give voice to those dispossessed by extraction. Thus, the knowledge projects associated with the real world of extractivism today ignore not only the degradation of land but the deterioration of relationships as well.

From these practices, we can tease out the metaphysical assumptions about value as well as the epistemic assumptions about knowledge. The sphere of value is circumscribed to only that which can be monetized and exchanged for profit. Profit is defined by what the extractor gains, without factoring in what others have lost. When profit is assumed to exhaust the sphere of value, a community's prior ideas about value may have been left unattended.

For the purposes of this chapter, then, I use the term *extractivism* to mean common practices of extracting monetized value that are linked to colonial histories and still embedded today in vastly unequal global economic and political power.

To name that which is extracted a *resource* and a potentially profit-making *value* is itself a substantive shift affecting our sense of ourselves and our relationality to the world. As Akeel Bilgrami (2016) notes, Gandhi raised these very questions about the language of European modernity. How did "nature" become "natural resources," Gandhi asked? How did our habitat become the focal point of projects of control and mastery? These were shifts in how we ontologize the world, but Gandhi also asked how the concept of knowledge to live by became transformed into a concept of expertise to rule by. This suggests the focal point of the next section: the epistemological orientation typical of extractivist projects.

Extractivism as an Epistemology

Extractivism as a practice conceptualizes the object of its pursuit, just as epistemologies conceptualize epistemic goals in varied ways.[1] When extractivism is motivated by the pursuit of commodifiable value, it tends to have certain metaphysical ideas about the nature of the value that it is pursuing.

First, as we'll see in the case studies, extractivist projects tend to assume that values can be identified in an objective and universal way.[2] Plural and competing conceptualizations of the value of an artifact or of a river or mountain challenge extractivist projects; thus, it becomes necessary to dismiss alternative views from other parties. When museum curators are seen as stewards who protect universal values, it becomes easier to accept the claim that they are the final arbiters of the value of Native American artifacts. Archaeologists likewise are often presented as working with superior epistemic methods, such as objectivity, traditionally defined, so that their assessments outrank others. Tribal group claims are often labeled particularistic and subjective, neither objective nor universal, and thus defeasible (Wylie 2002). They may also be defined as "resistance to science" (Fine-Dare 2002, 167). The important point to understand is that the process by which a value is defined is an interpretive practice for all parties. Allowing plural approaches to the definition of value disrupts the hegemony of academically trained experts. Thus, accepting pluralist approaches to value determination will involve accepting not only a pluralism of metaphysical commitments but also pluralist approaches to the knowledge processes that identify values subject to extraction.

Second, extractivist projects have an interest in defining the value that is extracted in a nonrelational way. This is a way of externalizing costs, as capitalists put it. The cost of cleaning up what has been made toxic in the process of extraction, the cost of the effects of extraction on infrastructure, the cost of regenerating raw materials when that is possible—none of these need to be addressed when the value of what is extracted can be portrayed as distinct and nonrelational (see, e.g., Wallerstein 2006, 57). Thus the assumption that values can be transported to a new domain without being diminished assumes a nonrelational, decontextualized conceptualization of values as discrete and independent of their relations to adjacent entities. Resources that are separated from their original context are not assumed to lose value in the process of removal unless they deteriorate in some material way. Further, those values that may be destroyed (such as arable land or animal habitats) in the process of extracting other values (such as mineral resources) are not codetermining or coconstitutive but merely adjacent and do not need to be factored into the quantification of the value achieved in the extraction. The belief that the values that have been destroyed in the process of extraction can be fairly paid for in a one-time payment also makes assumptions that values can be measured quantitatively and transformed into monetized elements that can then be calculated as a fair exchange.

In some cases, a value may not diminish in the process of extraction. But values can be diminished when, for example, artifacts are removed from their surroundings in ways that compromise the interpretation of their meaning, or plants with medicinal properties are separated from local communal practices that may impact their efficacy. The problem is when we assume without analysis that a nonrelational approach is adequate to assessing value. Rather, the relevance of context and relationality should be our initial assumption about the nature of value.

Thus far I have been discussing the metaphysical approach to the value of objects, but extractivists also assume that epistemic resources can be identified, extracted, and given a fair market value.[3] The key feature of an extractivist epistemology, I argue, is the way in which it treats this epistemic resource as separable from its origin, without subsequent loss, rendering it into a commodity with exchange value over which exclusive rights can be contractually defined, protected, and enforced. To be commodified in this way, knowledge must be reduced to information or data without involving interpretation or analysis, thus ensuring an inevitable epistemic injustice (Fricker 2007). When we understand knowledge as always involving interpretation, it becomes more obvious that dialogic processes are necessary for good epistemic outcomes.

Extractivist epistemologies thus involve certain epistemic assumptions about justification procedures that follow from their assumptions about the nature of the epistemic resources they are pursuing. Like extractivist capitalism, extractivist epistemologies attempt to extract epistemic elements from their original surroundings and in this way from their political, ethical, and institutional context of articulation. This makes it possible for knowledge-seeking institutions or individuals to avoid being held accountable to the ethical, political, and economic demands of indigenous groups or other local communities whose resources are being extracted; a one-time payment is sufficient to cover all obligations.

It may be the case that *knowledge extraction* is a mislabeling, since what is extracted is not identical to what existed before extraction. In general, extractivism denies its "making" relation to its object of pursuit, a relation avowed today by most trends in contemporary social epistemology and philosophy of science. The preferred language of *discovery* (rather than *making*) has ideological effects: *making* connotes craft, processes, and decisions, whereas the nonrelational characterization of *discoveries* may make dialogic approaches to knowing seem unnecessary, a political luxury without epistemic necessity. Extractors can perform identification, interpretation, hypothesis generation, analysis, judgment, and application on their own.

One of the epistemic questions we need to consider is whether all knowledge is extractable. To get at this question, it is useful to look at the claims of indigenous groups who have argued that some of the knowledge that extractors pursue is "inaccessible . . . untranslatable . . . unknowable" (Townsend-Gault, quoted in Valaskakis 2005, 84). As Valaskakis (Chippewa) points out, this is a "challenge to the power and privilege of outsiders." She cites Jimmie Durham, who, when asked about the meaning of a Cherokee text, replies, "I don't want them to know. . . . What I want them to know is that they can't know that."

There is an ambiguity in Durham's claim about the impossibility of knowing: Is it beyond the outsider's ability to know because he and others will not allow them access to the knowledge, or is it beyond their ability because the nature of the knowledge is such that it is lost in the process of extraction? We might imagine that extraction under conditions of colonialism renders outsiders truly bereft of the hermeneutic resources, and motivations, necessary to understand certain concepts and meanings. This is surely because epistemic collaborations require trust as well as a capacity to communicate effectively. As Vrinda Dalmiya (2016) has argued, successful knowing often requires attending to the qualitative relations between communities of knowers.

The failure to extract knowledge, then, may occur because those with insider knowledge refuse to cooperate, or because the differences in hermeneutic resources are such that outsiders cannot truly grasp the meanings even if they had cooperation. It follows that nondialogic and decontextualized conceptions of knowledge may compromise the quality of the extracted epistemic resource, affecting whether extraction can succeed to any extent, even partially.

As Dalmiya argues in *Caring to Know*, knowing is best achieved when one attends to the conditions and quality of the relations involved among knowers, and when all parties work to achieve trust. Given various forms of structural power imbalances, she argues that such trust is best achieved when dominant parties start from a position of relational humility. Western philosophy has long honored epistemic humility, and yet this humility is formulated in a nonrelational way as a feature simply about self-knowledge. I may know that I do not know, as Socrates put it, but believe that since I am aware of my ignorance, while others remain unaware of theirs, my humility motivates me to feel no need to seek others' counsel. Dalmiya's formulation of the epistemic norm of relational humility is quite different, as she explains: "What makes the humility relational is a connection between these two orientations where I am disposed to saying 'I do not know but you tell me where I have gone wrong.' Without the latter, we would merely have fallibilism and even skepticism: without the former we get a postmodern tendency toward epistemological pluralism [that is, relativism]. But with both in play, we end up with a healthy realism. . . . We foreground the epistemic authority of others while and in the act of acknowledging our own epistemic lacks" (2016, 119). Relational humility, then, unlike simple self-directed and nonrelational humility, motivates epistemic collaboration. Effective collaborations require taking an interest in the quality of the relations involved in specific projects of inquiry. Where Socratic humility is compatible with elitism, relational humility counsels a more egalitarian approach to knowing.

The critique of extractivist epistemologies provides a nonideal approach to the formulation of corrective epistemic norms. By taking into account existing nonideal conditions of knowing, we can develop norms of practice that might provide effective interventions with better outcomes. Relational humility as Dalmiya formulates it can provide us with our first corrective epistemic norm by avoiding the elitist implications of Socratic humility (what some might call false humility) with a socially aware relational humility. In the final section I discuss some issues about how to operationalize these norms across varying communities of knowers.

A second corrective epistemic norm would be to incorporate assessments of epistemic relations as a regular feature of the justificatory procedure. Truth claims are always relative to the evidence, and part of assessing the necessary evidence is an accounting of evidence gathering, interpretation, and assessment. Did knowers seek out and consider evidence that might dispute their favored hypothesis? Did knowers invite objections and consider the strongest objections? These standard norms of good scientific practice assume communities of knowers, but they need to address the power differentials that exist both within and between communities of knowers before they can determine whether the gathering and assessment of evidence was meaningfully rigorous. Thus, incorporating an assessment of relations directs knowers to consider how relations between knowers—both as individuals and as groups—affect the identification and assessment of evidence.

In the next section, I turn to two case studies of extractivist epistemological practices. The goal here is to address in particular the epistemic weaknesses of epistemological extractivism.

Case Studies

Delving into the real world of contested knowledge claims relating to extractivist projects reveals contrasting ideas about knowledge and knowers as well as methods of justification. Many of these contestations are local but have global significance. For example, the arguments over patenting food production processes affect how expertise is defined and whether the expertise of small farmers can compete with that of food engineers who work for multinational corporations.

The arguments over patents operate within the highly constrained discursive domain of courts, both local and international. Yet these domains make use of common ideas and concepts that reveal a variety of epistemic issues at play in the legal debates. By viewing the implicit epistemic norms operating in real-world struggles, we can begin to formulate corrective epistemic norms to address the operations of coloniality still at play. Some of these norms may need to be local, while others may have larger applicability, such as the need to review and assess the relationships between knowers.

Biopiracy and Patents

Our first case study considers biopiracy and the contemporary legal arguments over patenting plant material.

Transnational biotech companies have lately recognized the immense botanical knowledge of Amazonian peoples, among others, who have established continuous and thriving communities in rainforests where critical but rare plants grow, with important medicinal qualities. Yet capitalist approaches to this knowledge have largely been extractive rather than collaboratively relational (Escobar and Pardo 2007; Shiva 2020). Transnational pharmaceutical companies want to extract the knowledge from indigenous communities, even while the tribal populations are forcibly displaced so that agro-developers can gain access to their land, in some cases then becoming available labor for the companies engaged in extraction.

Conferring patents on natural organisms is useful for expanding the profitability of extractive procedures, but over the last several decades, patents have been opposed as a form of biopiracy. Vandana Shiva calls patents the heart of the new colonialism because they provide the means to establish unilateral control. She reminds us that the original patents were rights granted by sovereigns (Shiva 2007).

A certain amount of scientific knowledge is today accorded to indigenous groups, altering the "barbarian" projections of the past (see Castro-Gómez 2021, chap. 4). Yet there remain dramatic differences between indigenous groups and extractors about nature and land. Escobar and Pardo explore these differences under the rubrics of "constructions of nature," the "interpretive frame of biodiversity," and the requirements as well as the goals of "resource management." Many current battles over land and autonomy are now waged through claims over who controls the interpretive frames. To thwart "biopiracy" and "bio-imperialism," progressive NGOs call for "biodemocracy," which requires "local control of natural resources, the suspension of development mega-projects and subsidies for capitalist activities that destroy biodiversity . . . a redefinition of productivity and efficiency," and more (Escobar and Pardo 2007, 293).

What is at stake here is not just the interpretive frameworks of diverse groups but the epistemologies of diverse groups, or how they conceptualize knowledge and formulate justificatory processes. Inkeri Koskinen and Kristina Rolin describe "Scientific/Intellectual Movements," or SIMS, as collective efforts to devise and implement structural remedies that can address a series of epistemic injustices, both hermeneutic and testimonial. These SIMS are distinct from other social movements in having "knowledge-producing aims and not merely social and political aims" (Koskinen and Rolin 2019, 1057). In other words, SIMS contest existing normative epistemic practices in scientific projects. This requires putting epistemic concepts and norms of practice on the table for critique and reconstruction. Extractive capitalism encourages an unreflective epistemic practice that foreshortens the potential knowledge benefits that could accrue from more democratic, collaborative approaches that remain open to different epistemic practices.

The case of what has come to be called *biopiracy* will provide some examples of these points. Biopiracy is a concept meant to describe the assortment of activities by major transnational corporations to acquire exclusive rights over various resources and processes involving food and health. Something like twelve such corporations (constant mergers render the numbers unreliable, but it is small), mostly from the Global North, control the world's food and agriculture industries, which are often integrated with health industries that require the mass production of certain plants with pharmacological properties. These corporations use international and national laws to secure access to and power over what the world's population needs to survive. Along the way, entities of global capital affect how knowledge is defined and which epistemic resources come into view.

The colonization of regional economies requires controlling as much as possible every step in the production of necessary goods and extracting profits from every transaction. One way this has occurred is through the radical transformation and replacement of traditional seed production with a synthetic chemical process that does not regenerate and requires annual expenditures.

Corporate arguments for patent rights require characterizing how a given knowledge came into being as well as the ontological characterizations of certain kinds of entities. Living beings are transformed into raw material, and a forest becomes a field for extraction rather than a habitation to watch, nurture, and preserve from year to year (Garcia dos Santos 2007, 160–63; Code 2006). As Lorraine Code shows, farmers and indigenous communities often have distinct knowledge projects quite different than the knowledge projects of extractivists. When the goal is not extraction but habitation and annual farming, other sorts of knowledge become germane, requiring other sorts of epistemic activities. Observation and a limited experimentation that works with and maintains existing processes and rhythms of natural forest reproduction are quite different knowledge practices than the radical transformations involved in most extraction. Values are also redefined by extractivists as discrete and separable from their immersion in the holistic systems that are necessary for regeneration, so the extraction of values that diminish regeneration is not factored into the assessment of the overall value achieved.

When corporations claim to know better how to understand and make use of forest resources, to develop them efficiently and effectively to achieve the largest positive impact, they are covering over the fact that they are seeking different kinds of knowledge and may be redefining key terms. Part of the problem is that patent arguments do not allow for epistemic pluralism or multiple ways to define terms. Patent claimants must assert that the corporation has achieved an exclusive process and engaged in "invention," and that the profitability of this invention should be protected to encourage further innovation in these critical markets.

To be able to claim invention, claimants must pass two tests: a novelty test and an "obviousness" test. They must be able to claim that what they are patenting is new in some important respect and that it involves a procedure that is not obvious, such as simply scattering seeds of plants with known pesticidal properties around arable land. There are prior assumptions involved in how to discern what is new or obvious and enormous profits at stake, which may skew reasoning and inhibit transparency.

Those opposed to patent claims have argued that so-called new methods of extracting chemical properties are simply "straightforward applications of conventional organic chemistry and an extension of the traditional processes used for millennia" (Shiva 2020, 150). Claims about an application being "straightforward" or an "extension of traditional processes" are subject to numerous questions. Should an industrial reformulation of a traditional process be classified as new or as an extension? What counts as a "straightforward application" as opposed to an innovative application? And who gets to determine the relevant prior assumptions involved in defining these terms, and on what basis? All sides of the patent debates are interested parties, but considering the legitimacy of claims to either newness or obviousness requires the input of the parties with detailed knowledge of prior practices (see Whyte 2021).

The criterion involving obviousness is perhaps even more problematic than that of newness. Traditional processes are sometimes characterized as obvious although they are the product of extensive experimentation across generations. As Shiva argues, claims about the obviousness of certain agricultural practices may betray "a racist dismissal of indigenous knowledge . . . that evolved through extended and systematic knowledge development in non-western cultures" (2020, 150). Traditional approaches to farming are highly complex, involving methods of cultivating hybrids, a capacity for soil assessment, and the ability to assess how various crops affect the maintenance of soil health, as well as pest control and seed regeneration, among other things, all of which involve multiple communities and generations who share their experiences and knowledge. So questions of obviousness need to consider how what may seem to be an obvious process has been historically produced through collaborative empirical efforts.

Another term associated with biopiracy is "bio-prospecting," which, as Cori Hayden points out, is simply a new name for an old process of extracting usable knowledge (Greene 2002; Hayden 2003, 359). When such prospecting can yield patentable results with large profits, this can motivate some to engage in manipulations of the truth about the knowledge involved. The struggle over neem is one of the most well-known examples.

Neem is the name of a tree that flourishes in many parts of the Global South and has long been known about in South Asia: it was described in ancient Indian medicinal and religious texts, which gave detailed information about both its preparation and the ailments it could address. Its botanical name, *Azadirachta indica*, actually means "the free tree" (Shiva 2020, 145ff.).

Neem was not only made use of in the ancient world: in the twentieth century, a significant amount of research was done on the neem tree by the Indian Agriculture Research Institute as well as the Khadi and Village Industries Commission (Shiva 2020, 146). The central feature that began to draw attention in the developed world was neem's ability to provide pest control. In the 1970s, Western and Japanese companies became interested in the tree when growing opposition to chemical-based pesticides and other chemical products led to a search for natural, plant-based alternatives. This initiated the patent wars over neem. Starting in 1985, multiple patents for the production of synthetic forms were filed, and claimants had to devise ways to meet both the *new* and the *not obvious* criteria.

Those opposed to the patents argued that synthetic production was an extension of traditional processes, without sufficient innovation to be classified as new. For example, Eugene Schultz, chair of the National Research Council, held that "the biologically active polar chemicals could be extracted using technology already available to villages in developing countries" (quoted in Shiva 2020, 148). Opponents also argued that the newness claim was relative to location, and that patent offices in the West took up the neem claimant's cases because they were unfamiliar with the common knowledge of this plant in many non-Western areas of the world.

The effort to resist patenting neem involved a large coalition of research and policy groups as well as farming collectives and NGOs around the world, including the West. It took years of organizing with expensive legal teams to establish the fact that Western corporations had neither discovered the properties of neem nor invented the method of isolating the key causative elements of the tree.

We might say this is a species of the epistemology of ignorance in which the desire not to know about groups of knowers with competing or superior knowledge in certain domains is advanced through measures that enshrine a set of prior and questionable assumptions about how to operationalize terms such as *invention, discovery, new, obvious,* and *advanced.* Of course, it is not at all clear that patent defenders truly did not know about the prior knowledge in regard to neem or whether they were simply making strategic claims in the courts that the litigants did not necessarily believe. Yet the larger social effect of the language that patent claimants use can spread disinformation among wider publics.

To refer to my previous characterization of how extraction alters the ontology of the known, we can see in the example of the neem tree a recharacterization of knowledge, so that corporate claims of knowledge could be presented as if they were separable from their original context. In other words, it is likely that Western and Japanese interests studied the research and uses of neem prior to their efforts to develop new processes, but this history of knowledge had to be downplayed to establish the criteria international courts used to legitimate patents. However, setting the history aside helps to support the idea that dialogic models of knowing are unnecessary: that extractors can perform interpretation, analysis, judgment, and application unilaterally, that in collaborative processes the judgment of Western experts is final and determinative, and that their unilateral assessments can reach a threshold of epistemic justification fully adequate for a claim.

Neem is just one example, but the pattern it reveals of prospecting indigenous practices for possible patents and then using epistemic de-authorization as a means to secure the rights of extraction gives us some direction in developing a decolonial epistemology. More recently, different and more promising forms of collaboration around ethnobotany are developing across differently positioned groups. As Greene (2002) reports, these collaborations explode the idea of discovery, as if Westerners are working in virgin territory. They also challenge some prevalent ideas about traditional knowledge as unchanging and bounded to specific communities. Relational and dialogic approaches to knowing, then, will have an impact not only on the knowledge gained but on the concepts and assumptions that inform knowing practices and that motivate the effort to reform epistemic orientations that are imperialistic.

Museums as Knowledge Extractors

A second case study useful to explore here concerns the struggle over museum collections of artifacts and human remains.

Museums across the world have been collecting various objects from North American tribes for over a century, and millions of items are now housed in major institutions. In the United States alone, over 1,500 museums house several million such artifacts (Colwell 2017). Not surprisingly, the story of how these collections came into museum hands is troubling: sometimes items were purchased from their owners, but there was also outright theft, looting, and the plundering of graves on tribal lands.

Even those who purchased items legally from their owners did not always concern themselves with how these owners obtained their objects. In fact, early collectors often bragged about their predatory exploits. One of the most successful collectors was R. Stewart Culin, who prepared exhibitions for World's Fairs in 1892 and 1893 and provided over four thousand objects to the Brooklyn Museum still in its possession today. Culin specialized in Zuni artifacts and paid cash, explaining that "many of the Indians had nothing to eat" and that the money he gave them was used to purchase food (Colwell 2017, 17). One doubts that, under such circumstances, the Indians were able to bargain over prices.

Philip Deloria explains the epistemological outcomes of such extractions: "a collection created a vast web of possibilities for recontextualization, for moving objects out of one location . . . and into another. . . . The most important recontextualization may have centered on the authority of the collectors themselves, for the objects constituted them as unique figures of authority" (2018, 108).

The artifacts obtained include hundreds of thousands of humans remains, including those of both adults and children, and the funereal objects that had been placed in gravesites alongside them. Museums either put these remains on display for the public or held them in back storerooms for the exclusive access of researchers. This attitude toward human remains follows the practice of the British Museum, among others, who continue to display the bodies of ancient Egyptians, yet the display of indigenous people's remains concerns the recent victims of colonialist genocidal policies. Only some are identifiable as belonging to a particular tribe, and those not identifiable are classified as unaffiliated, creating special challenges to claims of rightful return.

For many decades now these practices of displaying human remains as well as the possessions of colonized peoples have been the subject of political and legal demands for return. As Valaskakis notes, "Along with land and treaty rights, Native people are laying claim to Indian objects and images, to museums and to history. . . . Across Indian Country, this move to transform the present and negotiate the future by recovering the past has contributed to new debates reclaiming memory, experience, and imagination" (2005, 81). She quotes an editor of the *Edmonton Journal,* who ruefully noted, "It has long been clear that we actually prefer our native culture in museums. We certainly do not prefer it running the Department of Indian Affairs. Nor do we prefer it announcing the news on national television or determining its own political destiny" (2005, 79). This makes it clear that museum displays concern group relations in the present and not merely the treatment of peoples from the past. Thus, we need to consider what specific kind of knowledge museums impart, and how they may impact the relations involved in the knowledge projects necessary for building new futures and improving group relations.

James Clifford (1988) referred to "salvage" forms of knowledge projects, particularly in anthropology and archaeology, where the science communities associated with the victors in a colonial struggle seek to salvage what is left of the "authentic" cultural remains of their victims. What is the purpose of such salvage? In some cases, individual collectors are motivated as trophy hunters, to display the fruit of conquest and give visible testament to settler victories. Museums in contrast offer more epistemic and universalist reasons: to protect significant sources of knowledge for the future of the global community. Material objects have long been considered more epistemically reliable than mere textual descriptions, and part of what motivated the creation of the field of archaeology was the belief that oral histories require some sort of material evidence. Ciriaco de' Pizzicolli, credited as the founder of archaeology, was an Italian merchant in the fourteenth century who fought in the Crusades. He argued that "ancient things" were more faithful sources of knowledge about classical antiquity than mere textual reports, even those written at the time (Fine-Dare 2002, 15). Given this view that material objects are essential for epistemic reliability, it can appear that Native groups who demand the return of objects and threaten lawsuits against institutions such as the Smithsonian National Museum of American History are simply antiscience.

In the 1960s, a coalition of tribes initiated a far-reaching demand for what came to be called *repatriation*: the return of the physical remains of the dead to their tribes for proper burial or other treatments deemed by their group to be appropriate and respectful (Mihesuah 2000). Museums resisted this with claims about scientific inquiry, historical interest, and larger public values, such as the ability to protect the remains more securely than the tribes could themselves (Jenkins 2016; Wylie 2002). The larger public values invoked included scientific, historical, and ethnographic research. It was argued that if museums were forced to repatriate all of these remains, significant sources of knowledge could be lost or compromised. By contrast, advocates for repatriation often made claims about the moral duty to respect those human beings whose bones were being bartered, monetized, and displayed. Some groups also made arguments, on the grounds of the powers of the dead over the living, that these remains could enact vengeance on those who mistreated them if they were not handled very carefully, and that proper burial rituals were required to assuage the anger of the dead. Adjudicating these conflicts required a choice about what kinds of reasons would be given priority or even acknowledged as valid.

In 1990, under increasing pressure from Native activists, the US government passed the Native American Graves Protection and Repatriation Act, known as NAGPRA. This law mandated that Native groups had the right to claim the remains of their tribal members. It was only museum officials, however, who were accorded the task of assessing whether there was sufficient evidence to identify the tribal association of specific remains, so the law did not instantiate real epistemic collaboration or evince a relational epistemic humility concerning the assessment of validity. As Deloria says, "NAGPRA . . . does not empower Indian people all that much" (2018, 112). In one consultative exchange process he witnessed, he concluded that "the museum, which recorded the discussions, gained far more from the exchange than it ended up giving to the tribe." Still, NAGPRA has had a major impact, leading to the return of tens of thousands of human remains. However, more than 100,000 continue to be housed in museums, along with millions of ceremonial objects used in funerals and burials such as beads, pipes, feathers, special clothing, ceramics, and other artistic products.

This is an ongoing struggle. European museums have no national laws requiring repatriations of this sort and have almost entirely refused to accede to demands from either nations or ethnic groups. In the United States, there are now efforts to strengthen the terms of NAGPRA, improve elements of the process to make it more collaborative, and extend its reach. Native groups are now demanding the return of unaffiliated remains that cannot be identified as having a relation to a specific tribe, asking that these still be removed from museum display cases even though they are not covered under NAGPRA. There is also contestation over what counts as the remains of living beings: Who determines the meaning of *living*? Some groups argue that artisanal or produced objects such as Ahayu:da, which are wooden objects taken to be war gods or "keepers of the sky," and that have long been traded as commodities among non-Native peoples, have spiritual elements and deserve to be included in repatriation laws despite the fact that tribal practices will then place these objects outside and allow them to deteriorate over time (Deloria 2018). Thus, the struggle is over ontological assumptions about how to conceptualize objects as well as how to prioritize diverse conceptions of value.

These debates continue to be represented as a conflict between the forces of reason, science, and modernity on the one hand, and on the other hand, premodern myth, superstition, and religion. This is a stacked deck in mainstream Western public discourses. Certainly, it is widely understood that the display of bodily remains, including scalps, is disrespectful, but this can be accorded without challenging existing ideas about knowledge.

Marisol de la Cadena (2015) recounts an interesting collaborative event between highly knowledgeable parties that still demonstrated an epistemic failure. The event involved an invitation to Quechua leader Nazario Turpo to be a consultant for an exhibition planned by the US National Museum of the American Indian (NMAI). The translation challenges in this collaboration were enormous, and, as she shows, the failures of translation occurred in a way of which the museum officials seemed unaware. Translation work across wide cultural divergences can only yield partial communication and cannot avoid some degree of misrepresentation. To refer to Turpo as a leader, as I just did, is itself a misrepresentation: his community's understanding of what it means to be *in-ayllu* is not representational but relational in ways that require some significant unpacking and dialogue for outsiders to grasp.

The NMAI pursued Turpo's help in a project of repatriation of human remains, but, as with the case of the Ahayu:da, there were challenges to the translation of concepts. Human remains are divided by the Quechua into two different sorts depending on the era in which they lived. Those that lived in a different era are referred to as *suq'akuna*, and their contact with living beings today is considered potentially dangerous. Turpo believed the remains in question may have lost their power or been "tamed" in the intervening period by changes in material conditions or the blessings of priests. Just to be safe, however, when the remains were returned, Turpo and his family organized several events to reduce the likelihood of negative effects. The NMAI officials, as de la Cadena recounts, "were thrilled to witness the ceremonies, which they saw as a celebration of the repatriation of ancestors' remains. The suq'akuna (as potentially dangerous entities) were lost in this translation of the event" (2015, 212).

De la Cadena describes this incident as one in a long series of mistranslations or, following Eduardo Viveiros de Castro (2004a, 2004b), "equivocations." An equivocation is neither an error nor a failure but a "communicative disjuncture . . . in which interlocutors are not talking about the same thing" (de la Cadena 2015, 27). Equivocations are constitutive features of cultural translation, thus unavoidable, but this also means that they are incorrectly understood as failures. Attempting to overcome them is the wrong goal, Viveiros de Castro argues. Colonial ventures sometimes go so far as to destroy languages in a vain attempt to produce singular, unified meanings; epistemic collaborations are hobbled by such methods, not enhanced.

A dialogic collaboration across cultures cannot be effective if it seeks to purge all differences in the pursuit of complete univocality. Instead, there needs to be a steady awareness of divergences of meaning within communicative practices. Mutually productive relations between speakers do not require perfect translatability and in fact, the attempt to purge differences will likely diminish relations, erode trust, and adversely affect what understanding is genuinely possible. If equivocations are taken to be failures, this could motivate silencing and the downplaying of divergences.

The case studies just discussed yield lessons about the move to a relation-based, dialogic practice of knowing, but also about the challenges to creating collaborative projects. Let me summarize these before moving on to the final section.

The case of neem showcases the necessarily collaborative nature of human knowledge. Farmers could not develop techniques of cultivation if they refused to share their knowledge with outsiders or future generations. The use of patents to hoard knowledge and to thwart cooperation between the complex knowledge communities that can contribute to making the most out of the critical causative properties of given plants has a negative epistemic impact: not only promoting the misrepresentation of practices as new or obvious but also curtailing opportunities for expansion of knowledge through collaboration. Patents create distrust and erode potential relationships among knowers. The principal motivation for private patents is private profit, but this is not the only motivation human beings have to innovate. Nor is private gain the best or most efficient route to innovation and development; in fact, these are likely to be stymied by exclusionary rules that curtail creativity and open communication.

The problem with museums that house human remains cannot be represented as a moral and political problem without epistemic implications. Material objects are sources of knowledge, as de' Pizzicolli claimed, and yet their interpretation requires contextualization and collaboration with cultural insiders. Dialogically pursued projects cannot epistemically rank the contributions of various parties in an a priori way, so that only one side is assumed to have the capacity to judge claims or understand significance. Instead, collaborators should be open to learning from anyone. The conditions necessary for fruitful collaborations should be explored and protected, even while the inevitability of equivocation, in Viveiros de Castro's sense, is accepted. If the focus turns to developing robust relations over time rather than securing particular outcomes that can be monetized, the possibilities for advancing greater understanding will be enhanced.

Extractivist epistemologies that have developed in the epoch of global empires have specific features that facilitate the extraction of knowledge from subaltern groups to dominant groups. Such facilitation is enhanced by specific and concrete practices but also by epistemic ideas that conceptualize knowledge and justification in ways that legitimate noncooperation and excuse nontransparency. For example, oversimplified conceptions of expertise downplay the role of interpretation and the inevitably perspectival aspect of knowing, as we've seen in the two case studies just discussed.

Clearly, profit making can undermine the motive to collaborate or to be transparent. But beyond assessing the epistemic assumptions involved in specific profit-making projects, I argue we need to consider how the ideas and practices that have undergirded colonial and capitalist practices of extraction have influenced general approaches to knowledge. It may well be that extractivism has provided a model of knowing with a reach far beyond mining projects or biopiracy. As we've seen, museum practices may be genuinely motivated by an interest in historical truth rather than profit yet have long felt justified in operating without collaboration with "nonexperts." This larger claim about the effects of colonialism on Western epistemologies is a task for historians of ideas, but epistemologists and philosophers of science are beginning to weigh in on the social contexts that have affected which ideas become influential (e.g., Tuana 1992; Tiles and Tiles 1993; Dussel 1995; Potter 2001; Harding 2008).

We can now develop a more precise definition of extractivist epistemologies as exhibiting four features: the practice of ranking knowers, denying the need for collaboration across groups, defining values as nonrelational and objectively determinable, and seeking exclusive appropriation and control over intellectual items such as knowledge and processes. Defining extractivist epistemologies with these four points excludes collaborative projects in which one group may seek knowledge from another but without seeking exclusive control or ranking knowers.

In this final section, I elaborate four corrective epistemic norms that can counteract extractivist epistemologies: (1) acknowledging the incompleteness of all knowledge, (2) developing an approach that recognizes plural epistemologies and seeks productive relationships of inter-epistemology, (3) practicing relational epistemic humility, and (4) regularizing the assessment of epistemic relationships in projects of knowing.

Boaventura de Sousa Santos calls for a principle of the incompleteness of all knowledge, meaning that no singular approach to knowing will ever achieve absoluteness or sufficiency. Sciences and technologies are developed in historically contingent ways with concepts and practices that are affected by contextual conditions. As we've seen, museum curators trained in Western methods and tribal members may operate with quite divergent ontologies, yet both can reveal some aspects of a given artifact: curators can often date objects, while tribal members can provide context and history, but we should note that the usage of concepts like *object*, *artifact*, and so on may not be common across these different approaches. De Sousa Santos argues that, instead of seeking a singular approach, we should acknowledge that "there is no essential or definitive way of describing, ordering, and classifying processes, entities, and relationships in the world" (2014, 196). In relation to specific goals, there will be better and worse ways to describe, order, and classify phenomena, but this does not entail a singular approach that will someday win out overall, or that aiming for a singular approach is the most fruitful way to pursue knowledge. It may disable the development and improvement of any specific approach to knowing practices (see also Massimi 2022).

The claim of incompleteness is familiar in twentieth-century traditions of philosophy of science, especially the historicist trend developed by Imre Lakatos, Thomas Kuhn, Hilary Putnam, Paul Feyerabend, and others. As Putnam (1981) argues, the impressive achievements of modern science are compatible with the partiality of their ontological picture of the world, since Western ontologies are based on contingent conceptual repertoires that will no doubt be modified if not replaced (see also Boyd 1989). Hence, incompleteness does not challenge all versions of scientific realism. And acknowledging incompleteness motivates engagement with other approaches that may operate with quite different concepts, and it counters the disposition to engage in an overall ranking of approaches at some contrived metalevel. Thus, starting from an acceptance of incompleteness can serve as a corrective norm by motivating an openness to divergent ideas and practices.[4]

The second corrective norm follows from the first. If we acknowledge incompleteness, we need to develop a concept of "inter-epistemology" as an alternative to epistemic imperialism (de Sousa Santos 2014, 2018). The stance of an inter-epistemology is to accept that diverse approaches to knowing may be epistemically valid or productive, that any given approach will be incomplete, and thus we need to think through the terms by which we can develop constructive relationships between different approaches. This need not be a relativism that would accommodate all theories of knowledge or disallow the raising of critical questions across different approaches, but it would reject the idea that the only intelligible epistemological goal is a unified theory that will overcome all differences in how knowledge is defined and how knowing practices are pursued. As de Sousa Santos explains, "In the ecology of knowledge, finding credibility for nonscientific knowledge does not entail discrediting scientific knowledge. It implies, rather, using it in a broader context of dialogue with another knowledge" (2018, 189). Thus, the concept of inter-epistemology takes us a step further than merely acknowledging incompleteness, because it also acknowledges the existence of fundamental diversities in conceptualizing knowledge.

Mignolo has argued against the use of the term *epistemology* itself for fear that the weight of its typical usage in the modern West carries framing assumptions that deny the interpretive element in knowing. Richard Rorty (1979) held a similar view and argued for the term *hermeneutics* as a replacement. Modern Western epistemologies tend to seek a systematic or metalevel approach that judges the doxastic content of common sense as if from above, without an interpretive frame. Mignolo (2000, 9) favors the older concept of gnoseology, which, he argues, takes a more general approach to knowledge, without privileging science or assuming there is an essential form of valid knowing that rules all practices (see also Alcoff 2007). He wants a concept that combines what Western modernity separated, that is, epistemology and hermeneutics, or knowledge and the interpretive frameworks of a given cultural location, so that we can move to a pluralist approach.

De Sousa Santos's usage of the term *inter-epistemology*, however, unites with this project. It is an effort to bring epistemology down from the clouds so that it can accept multiple framing assumptions. And as previously stated, influential trends in twentieth-century Western epistemology and philosophy of science, such as pragmatism, historicism, naturalized epistemology, and contextualism, can accord with Mignolo's aims as well. But Mignolo's critique helps to shift our focus to the constitutive nature of interpretive elements in our practices and the need for a pluralist stance. The question is, how do we enhance practices of inter-epistemological collaboration?

We need to explore, more than can be done in one chapter, the ideas that inhibit collaboration across epistemological differences. One of the typical criticisms of (so-called) traditional approaches to knowing is the worry about conformism. Traditional approaches generally do not work on individualist models of knowing, in which individualism is assumed to be the generator of advances in knowing. Knowledge achieved over millennia operates with a respect for the past and a diffuse understanding of invention and ownership. Who owns the traditional process of cultivating the properties of the neem tree? This diffused framing of invention is a more realistic account of the nature of intellectual work in general, but these approaches may be seen as having endemic conformist tendencies that can limit innovation. De Sousa Santos quotes the didactic sage Odera Oruka, who reports, however, that popular wisdom can be subjected to debate, as well as "the communal set-up" that sometimes tries to fend off challenges. And it is also important to recall that the same dangers of a dysfunctional conformism can be found in Western sciences, as many have argued (see Lloyd [2006] and Smolin [2006] for recent case studies). No system of knowledge may be immune from the danger of conformism, and remaining open to alternative epistemologies can provide a helpfully corrective norm to ward off calcification.

Another challenge to the idea of inter-epistemology is the assumption that multiple research communities can only cooperate if all participants share methodological approaches and standards. But this requirement can inhibit collaboration before it is begun and may reduce its epistemic value. To see why, we need to understand that methodological standards are not objective but "conventional in character and irreducibly social" (Wilholt 2016, 223; see also Rudner 1953; Douglas 2009).

Research projects are guided by specific aims, such as whether to value positive results over negative results or vice versa. These may work at cross-purposes, so both cannot be valued equally. Rapid HIV tests and certain COVID tests confer highly reliable negative results but highly unreliable positive results. Methodological choices are thus guided by priorities of utility as well as decisions about resource allocation. The impact of false positives is generally lower from a public health perspective than the impact of false negatives, especially if a positive result from a rapid test can be quickly followed up with a more reliable test.

The epistemic advantage of collaboration across diverse communities with divergent priorities thus becomes clear: when standards and methods come out from behind the cloak of (traditional) objectivity and disinterestedness, debates must include a recognition that diverse goals can shift assessments of practices. It then becomes clear why collaborations need not pursue consensus on the research methods. Dissensus about priorities and values may enlarge the overall understanding of the phenomena, such as when human remains are viewed divergently as (1) a source of cultural knowledge that overrides all other considerations, (2) entities with intrinsic moral value that requires respect and deference, or (3) potentially dangerous forces. Each of these counsel different methods of engagement with human remains. Collaboration does not require that only one understanding wins out: there may be ways to accommodate more than one goal and more than one method of engagement. Thus, if the diversity in understanding human remains is viewed as an inevitable impediment to knowledge, epistemic outcomes will be unnecessarily foreshortened.

Yet a global relativism should also be rejected because some practices of knowing do not interact fruitfully but degrade trust and disable motivations to cooperate. As Sandra Harding explains, "only modern Western sciences" are viewed as having "the resources to escape the universal human tendency to project onto nature cultural assumptions" (2008, 4). When Western science is viewed as exceptional and triumphal over all others, Harding explains, relationships are disabled. Such frameworks portray a zero-sum game in regard to contrasting epistemologies and methodologies. Moreover, triumphalism and exceptionalism inhibit critique from the outside, preempting the potential benefits of openness to criticism built into the scientific method. De Sousa Santos suggests this is the often-unacknowledged defining feature of hegemony-seeking epistemologies, that "they only recognize internal limits" (2018, 189). The concept of inter-epistemology aims instead to approach epistemological diversity as a potential epistemic resource for all parties even if no singular epistemology achieves consensus.

A third corrective norm is the one discussed earlier as developed by Dalmiya: relational humility. The history of the pursuit of knowledge under the *longue durée* of colonialism is reason enough to seek to develop an unfamiliar humility. Abrupt dismissals of other claims to know have not served the West very well; many are now trying to develop sustainable practices for natural environments, for example, and recognize that Western science has much to learn in this area from indigenous knowers. But there is also a more general methodological reason to cultivate relational humility.

The neem case is particularly instructive. Profit-making motivations can work against humility by simply aiming for a legal triumph rather than expanding knowledge and understanding. Achieving the right to exclusive patents may tempt litigants to conceal information and portray indigenous farmers as ignorant and backward, thus enabling the continuance of racist and sexist ideas. Adversarial court systems provide little motivation for relational humility. In truth, the enormous organizational efforts and financial costs incurred by holding off neem patents should never have been required: the international court system should not allow private corporations to initiate claims without establishing that they have worked collaboratively with the relevant knowers before deciding to push their patent requests. To make genuine collaboration necessary would impact the power imbalance, giving locals an effective veto power.

Relational humility is a critically important corrective to centuries of colonialism that still reverberate in our discourses and institutions. As Dalmiya argues, in this context relational humility is not only a moral virtue but also, as she says, "an epistemic *excellence*" (2016, 115, emphasis in original). The point is not simply to ascribe ignorance to oneself, but also to ascribe knowledge, or likely knowledge, to others. This entails changes in practice but also an altered self-understanding (Dalmiya 2016, 122). The positive effects of relational humility are both moral and epistemic, since it allows those with higher status "to learn from people at the epistemic and social margins" (97). The neem case demonstrates the lack of relational humility not only toward small farmers but also toward the Indian Agriculture Research Institute and the Khadi and Village Industries Commission that had already done extensive research on the neem tree.

Many Western sciences, as Harding and others argue, have long promoted the idea that knowledge is enhanced by individualism, arrogance, and making a sharp division between moral and epistemic issues of concern. The concept of the genius often portrays a lone figure who makes his path, ignoring its repercussions for others. The significance of Dalmiya's concept of relational humility is that it shows the weaknesses in these arguments from an epistemic point of view. Standard conceptions of objectivity counsel openness to criticisms, or the ability to consider every possible objection to one's favored view, but if the viability of objections is assessed by only one side, there may not be "strong objectivity" as Harding (2015) calls for. The practice of preemptively ranking knowers offers a way to deflect their alternative views based on spurious considerations, such as illiteracy or lack of formal education.

However, relational humility is not an attitude that is exclusively relevant to the dominant parties or those from colonizing nations. It is always possible that others know things that I do not know. Overturning histories of domination requires formerly subordinate groups to reflect on our tendency to lack confidence and manifest excessive deference and humility, but we should define the problem as excessive humility, not humility per se, since the sort of epistemic arrogance that assumes I have nothing to learn from others, all sorts of others, will not serve my own epistemic and liberatory aims.

Work on the epistemologies of the social sciences has long discussed the idea of insider and outsider knowledge. Insiders to a sphere of practice, a culture, a language, and so on, will have clear epistemic advantages in picking up nuances and distinctions that others may miss. This idea has been marshaled against the claim that the most reliable knower will be an objective outsider, without attachments or investments of any sort. The reality is that all knowers are positioned in some relationship to the knowledge project, and these include such things as profit and individual career advancement. Hence all knowers have investments of some sort. Still, it makes sense to distinguish insider and outsider positions not on the grounds of disinterestedness but in regard to the specific relationship to the knowledge being pursued. And it also makes sense to see both positionalities as having potential epistemic advantages. As Edward Said (2004) pointed out, although they are in some ways outsiders, colonial subjects have shed new light on the great literature of Europe, raising new questions and developing new interpretations. Outsiders have detected subtle patterns of self-protection and racist myths operating in Western literature and philosophy. The lesson is that neither outsider nor insider viewpoints have a priori privilege. Diverse positionalities can increase the interpretive frames and thus enlarge understanding.

The fourth and final corrective norm that can redress extractivist epistemologies is to require an analysis of the qualities of relationships between the relevant parties to a knowledge project as part of standard procedure. Shifting to the stance of inter-epistemology and giving up universalist aspirations involves developing our thinking about what kind of relationships across epistemological differences are possible as well as functional. Pursuing a relational humility requires a conscious reorientation of our epistemic prejudgments of others, but this needs to be built into institutional mandates rather than left to individuals.

Attending to the quality of our relationships should involve an examination of trust, reciprocity, care, lack of envy, patience, and benevolence. While I have been emphasizing the epistemic reasons that should motivate our move away from extractivism, it is important to acknowledge that, in truth, epistemic considerations are bound up with other considerations. The social character of research, its reliance on conventions and local assumptions, the variability of interpretation and judgment, all point to the need to consider the normative character of social relationships between diverse groups that are situated differently vis-à-vis political and economic power. Despite inequality, all sides may benefit from becoming aware of multiple and conflicting assumptions and priorities: these can enhance group self-awareness and the possibility of an eventual shared understanding, even though this is likely to remain partial.

Attending to the moral and political character of relationships is thus critical for knowledge, even if, as Dalmiya perceptively points out, enhancing relationships requires shutting down some knowledge projects or lines of inquiry. This is not a paradox but a feature of long-term epistemic relationships. Museum curators must be prepared to relinquish some epistemic goals to achieve productive relationships that include a sense of trust and care. These values must be manifested in practice and not merely in word, and this requires forgoing some pursuits.

To conclude, to overcome extractivist influences in our norms and practices of knowing, we need to seek more genuinely egalitarian epistemic collaborations without presumptively discrediting certain knowers or alternative epistemologies.

Notes

This chapter appeared originally as an article in *Tapuya: Latin American Science, Technology and Society* 5, no. 1 (2022), https://doi.org/10.1080/25729861.2022.2127231.

1   The term *epistemology* is sometimes used very loosely these days. For the purposes of this chapter, I define epistemology more narrowly to focus on theoretical analyses of knowledge, such as what counts as knowledge and knowing practices. But I take a broader approach than the usual Western focus on overcoming skepticism or the search for universal justificatory norms (see Alcoff 1996, 4). Especially within decolonial work, epistemology should remain a broad umbrella, but it can continue to be a term that signals a focus on the ideas about why certain practices have better epistemic outcomes than others, however those outcomes may be variously defined. I discuss this choice of terms further in the last section.

2   Just to be clear, here I am referring to *values* only in the Marxist sense as something that has either use-value or exchange-value or both. This is not the same sense of the word used in the idea that epistemic justification is guided by normative value commitments.

3   Of course, all of this terminology is subject to contestation: not everyone views mountains or rivers as objects separable from human beings, and the concepts of metaphysics and epistemology carry much Western baggage. This reveals the difficulty of overcoming cognitive imperialism and achieving the inter-epistemology that de Sousa Santos calls for. I discuss this further in the last section, but I follow the approach of Viveiros de Castro in rejecting the dream of a perfect translation or of a metalevel language that can adjudicate all semantic differences.

4   One might think that these corrective norms are operating at a metalevel, but I am putting these forward as correctives for the particular shortcomings of many current Western approaches to knowing, though of course they may be helpful elsewhere.

References

Alcoff, Linda Martín. 1996. *Real Knowing: New Versions of the Coherence Theory*. Ithaca, NY: Cornell University Press.

Alcoff, Linda Martín. 2007. "Mignolo's Epistemology of Coloniality." *New Centennial Review* 7 (3): 79–102.

Bebbington, Anthony. 2010. "Extractive Industries and Stunted States: Conflict, Responsibility and Institutional Change in the Andes." In *Corporate Social Responsibility: Discourses, Practices, Perspectives*, edited by K. Ravi Raman, 97–115. London: Palgrave Macmillan.

Bilgrami, Akeel. 2016. "Gandhi's Radicalism: An Interpretation." In *Beyond the Secular West*, edited by Akeel Bilgrami, 215–45. New York: Columbia University Press.

Boyd, Richard. 1989. "What Realism Implies and What It Does Not." *Dialectica* 43 (1–2): 5–29.

Castro-Gómez, Santiago. 2021. *Zero-Point Hubris: Science, Race, and Enlightenment in Eighteenth-Century Latin America*. Translated by George Ciccariello-Maher and Don T. Deere. New York: Rowman and Littlefield.

Clifford, James. 1988. *The Predicament of Culture: Twentieth-Century Ethnography*. Boston: Harvard University Press.

Code, Lorraine. 2006. *Ecological Thinking: The Politics of Epistemic Location*. New York: Oxford University Press.

Colwell, Chip. 2017. *Plundered Skulls and Stolen Spirits: Inside the Fight to Reclaim Native America's Culture*. Chicago: University of Chicago Press.

Dalmiya, Vrinda. 2016. *Caring to Know: Comparative Care Ethics, Feminist Epistemology and the Mahabharata*. New York: Oxford University Press.

de la Cadena, Marisol. 2015. *Earth Beings: Ecologies of Practice across Andean Worlds*. Durham, NC: Duke University Press.

Deloria, Philip J. 2018. "The New World of the Indigenous Museum." *Daedalus* 147 (2): 106–15.

de Sousa Santos, Boaventura. 2014. *Epistemologies of the South: Justice against Epistemicide*. New York: Routledge.

de Sousa Santos, Boaventura. 2018. *The End of the Cognitive Empire: The Coming of Age of Epistemologies of the South*. Durham, NC: Duke University Press.

Douglas, Heather. 2009. *Science, Policy, and the Value-Free Ideal*. Pittsburgh, PA: University of Pittsburgh Press.

Dussel, Enrique. 1995. *The Invention of the Americas: Eclipse of "the Other" and the Myth of Modernity*. Translated by Michael D. Barber. New York: Continuum.

Escobar, Arturo, and Mauricio Pardo. 2007. "Social Movements and Biodiversity on the Pacific Coast of Colombia." In *Another Knowledge Is Possible: Beyond Northern Epistemologies*, edited by Boaventura de Sousa Santos, 288–314. New York: Verso.

Fine-Dare, Kathleen S. 2002. *Grave Injustice: The American Indian Repatriation Movement and NAGPRA*. Lincoln: University of Nebraska Press.

Fricker, Miranda. 2007. *Epistemic Injustice: Power and the Ethics of Knowing*. Oxford: Oxford University Press.

Garcia dos Santos, Laymert. 2007. "High-Tech Plundering, Biodiversity, and Cultural Erosion: The Case of Brazil." In *Another Knowledge Is Possible: Beyond Northern Epistemologies*, edited by Boaventura de Sousa Santos, 151–81. New York: Verso.

Greene, Shane. 2002. "Intellectual Property, Resources or Territory? Reframing the Debate over Indigenous Rights, Traditional Knowledge, and Pharmaceutical Bioprospection." In *Truth Claims: Representation and Human Rights*, edited by Mark Bradley and Patrice Petro, 229–49. New Brunswick, NJ: Rutgers University Press.

Harding, Sandra. 2008. *Sciences from Below: Feminisms, Postcolonialities, and Modernities*. Durham, NC: Duke University Press.

Harding, Sandra. 2015. *Objectivity and Diversity: Another Logic of Scientific Research*. Chicago: University of Chicago Press.

Harvey, David. 2003. *The New Imperialism*. Oxford: Oxford University Press.

Harvey, David. 2004. "The 'New' Imperialism: Accumulation by Dispossession." *Socialist Register* 40: 63–87.

Hayden, Cori. 2003. "From Market to Market: Bioprospecting's Idioms of Inclusion." *American Ethnologist* 30 (3): 359–71.

Jenkins, Tiffany. 2016. *Keeping Their Marbles: How the Treasures of the Past Ended Up in Museums . . . and Why They Should Stay There*. New York: Oxford University Press.

Khader, Serene. 2019. *Decolonizing Universalism: A Transnational Feminist Ethic*. New York: Oxford University Press.

Koskinen, Inkeri, and Kristina Rolin. 2019. "Scientific/Intellectual Movements Remedying Epistemic Injustice: The Case of Indigenous Studies." *Philosophy of Science* 86 (5): 1052–63.

Lloyd, Elisabeth A. 2006. *The Case of the Female Orgasm: Bias in the Science of Evolution*. Boston: Harvard University Press.

Massimi, Michela. 2022. *Perspectival Realism*. New York: Oxford University Press.

Mignolo, Walter D. 2000. *Local Histories/Global Designs: Coloniality, Subaltern Knowledges, and Border Thinking*. Princeton, NJ: Princeton University Press.

Mihesuah, Devon A., ed. 2000. *Repatriation: Who Owns American Indian Remains?* Lincoln: University of Nebraska Press.

Mills, Charles. 2005. "'Ideal Theory' as Ideology." *Hypatia* 10 (3): 165–84.

Petras, James. 2012. "Extractive Capitalism and the Divisions in the Latin American Progressive Camp." *Global Research*, May 3, 2012.

Potter, Elizabeth. 2001. *Gender and Boyle's Law of Gases*. Bloomington: Indiana University Press.

Putnam, Hilary. 1981. *Reason, Truth and History*. Cambridge: Cambridge University Press.

Riofrancos, Thea. 2020. *Resource Radicals: From Petro-nationalism to Postextractivism in Ecuador*. Durham, NC: Duke University Press.

Robinson, William I. 2008. *Latin America and Global Capitalism: A Critical Globalization Perspective*. Baltimore, MD: Johns Hopkins University Press.

Rorty, Richard. 1979. *Philosophy and the Mirror of Nature*. Princeton, NJ: Princeton University Press.

Rudner, Richard. 1953. "The Scientist qua Scientist Makes Value Judgments." *Philosophy of Science* 20 (1): 1–6.

Said, Edward. 2004. *Humanism and Democratic Criticism*. New York: Columbia University Press.

Sankey, Kyla. 2014. "Colombia: The Mining Boom: A Catalyst of Development or Resistance?" In *The New Extractivism: A Post-neoliberal Development Model or Imperialism of the Twenty-First Century?*, edited by Henry Veltmeyer and James Petras, 114–43. New York: Zed.

Shiva, Vandana. 2007. "Biodiversity, Intellectual Property Rights, and Globalization." In *Another Knowledge Is Possible: Beyond Northern Epistemologies*, edited by Boaventura de Sousa Santos, 272–87. New York: Verso.

Shiva, Vandana. 2020. *Reclaiming the Commons: Biodiversity, Indigenous Knowledge, and the Rights of Mother Earth*. Santa Fe, NM: Synergetic Press.

Smolin, Lee. 2006. *The Trouble with Physics: The Rise of String Theory, the Fall of a Science, and What Comes Next*. New York: Houghton, Mifflin, Harcourt.

Tiles, Mary, and Jim Tiles. 1993. *An Introduction to Historical Epistemology: The Authority of Knowledge*. Oxford: Blackwell.

Tuana, Nancy. 1992. *Woman and the History of Philosophy*. New York: Paragon House.

Valaskakis, Gail Guthrie. 2005. *Indian Country: Essays on Contemporary Native Culture*. Waterloo, ON: Wilfred Laurier University Press.

Veltmeyer, Henry, and James Petras, eds. 2014. *The New Extractivism: A Postneoliberal Development Model or Imperialism of the Twenty-First Century?* New York: Zed.

Viveiros de Castro, Eduardo. 2004a. "Exchanging Perspectives: The Transformation of Objects into Subjects in Amerindian Ontologies." *Common Knowledge* 25 (1–3): 21–42.

Viveiros de Castro, Eduardo. 2004b. "Perspectival Anthropology and the Method of Controlled Equivocation." *Tipití* 2 (1): 3–22.

Wallerstein, Immanuel. 2006. *European Universalism: The Rhetoric of Power.* New York: New Press.

Whyte, Kyle. 2021. "Against Crisis Epistemology." In *Routledge Handbook of Critical Indigenous Studies,* edited by Brendan Hokowhitu, A. Moreton-Robinson, L. Tuhiwai-Smith, C. Andersen, and S. Larkin, 52–64. New York: Routledge.

Wilholt, Torsten. 2016. "Collaborative Research, Scientific Communities, and the Social Diffusion of Trustworthiness." In *The Epistemic Life of Groups: Essays in the Epistemology of Collectives,* edited by Michael S. Brady and Miranda Fricker, 218–34. New York: Oxford University Press.

World Bank. 2005. *Extractive Industries and Sustainable Development: An Evaluation of World Bank Group Experience.* Washington, DC: World Bank / International Finance Corporation / Multilateral Investment Guarantee Agency.

Wylie, Alison. 2002. *Thinking from Things: Essays in the Philosophy of Archaeology.* Berkeley: University of California Press.

Epistemic Decentralizing

Revisiting Knowledge
Asymmetries from the Periphery

In this chapter, we advance the program outlined in the introduction by exploring the epistemic consequences of separating, for analytical reasons, the decentering of ideas from the decentralization of infrastructures and practices. Initially, I describe four scenarios that are produced by linking two variables that have tended to be merged in the literature: the decentering and decentralization of knowledge. This gives rise to (1) epistemic indifference, (2) epistemic extractivism, (3) epistemic co-optation, and (4) epistemic mutualism. Going deeper, we will see that academic mutualism, an outcome of epistemic humility (Dalmiya 2016; Grosfoguel 2022; chapter 1, this volume), is a condition for achieving high levels of epistemic decentralization and highly decentered knowledge. Subsequently, I analyze two Latin American cases in which the process of epistemic decentralization can be observed: the restitution of Inca remains to the Peruvian government by Yale University and the transformation of Chilean astronomy through the installation and operation of a state-of-the-art telescope in the Atacama Desert. Through the examples, it is possible to observe certain stages of the process, but also to appreciate its limitations—that is, the paths that remain to be traveled. The chapter ends with reflections on the empowerment of actors through epistemic decentralization. Their agency as cognitive subjects is, to a large extent, the result of academic fields moving, gradually but steadily, toward forms of epistemic mutualism.

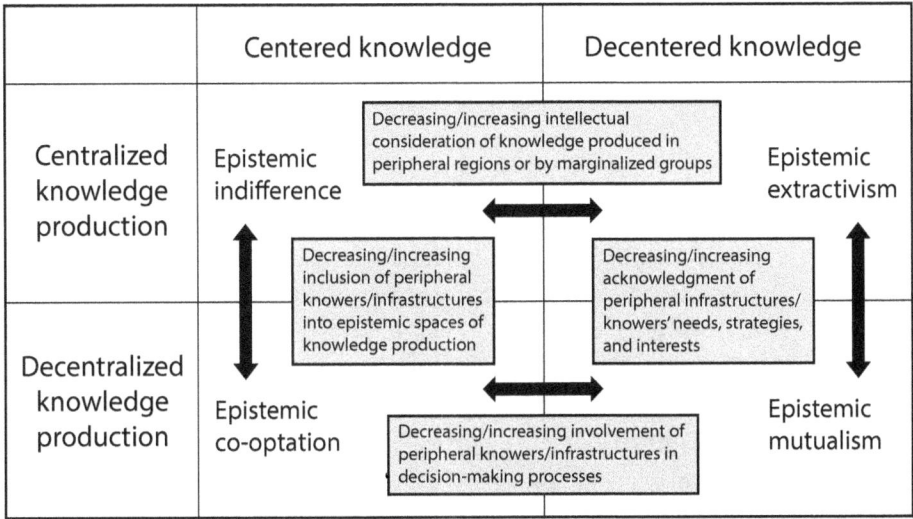

	Centered knowledge	Decentered knowledge
Centralized knowledge production	Epistemic indifference	Epistemic extractivism
Decentralized knowledge production	Epistemic co-optation	Epistemic mutualism

Decreasing/increasing intellectual consideration of knowledge produced in peripheral regions or by marginalized groups

Decreasing/increasing inclusion of peripheral knowers/infrastructures into epistemic spaces of knowledge production

Decreasing/increasing acknowledgment of peripheral infrastructures/ knowers' needs, strategies, and interests

Decreasing/increasing involvement of peripheral knowers/infrastructures in decision-making processes

Figure 2.1 Stages in the cycle of (de)centering and (de)centralizing.

(De)centering and (De)centralizing: Dynamics and Narratives

In this section I do two things simultaneously. First, I tell stories, drawing on accounts told to me by Latin American scholars over more than fifteen years of my research on the production and circulation of knowledge in peripheral areas. I want to point out that what was often counted as a minor issue or difficult for me to conceptualize in the light of current theories now takes on new importance. These fragments make it possible to observe tendencies whose understanding will depend on delving deeper into the processes of (de)centering and (de)centralizing. For this reason, parallel to the stories, there will be a development of the four scenarios that emerge from the intertwining of these two processes. I call these epistemic indifference, epistemic extractivism, epistemic co-optation, and epistemic mutualism, as shown in figure 2.1.

Epistemic Indifference

We don't use a lot of English . . . and sometimes we do things that have a local importance. So that article is not so publishable. Of what you produce, little of it is likely to appear in the indexes. In *Sociological Abstracts*

I have two or three things, out of a mass of almost 45 papers. . . . I'm telling you that many colleagues don't have a single one! We are very out of touch with the world . . . besides the fact that you come from the end of the world.

—ARGENTINE SOCIOLOGIST

What is disturbing and really annoying is when I go to the United States, because American professors believe that the academic world and subjects like this can be approached by reading 90% of the texts in English. Man! If the subject is the Hispanic world, the political and intellectual history of the first quarter of the 19th century, it's common sense that you have to read a lot in Spanish, right? On top of that, the best academics write in Spanish!

—MEXICAN HISTORIAN

About four years ago I invited an American colleague and his initial answer was yes. Then he asked me: "And what do you want me to present?" I replied, "I think it would be interesting to have a general reflection on the importance of a focus on stratification, on contemporary social mobility, on the analysis of Latin American cities." Then he answered me almost immediately, "You know what, I'm not going. Those kinds of generalities don't interest me." He cut me off. A lot of them are on their wavelength and very interested in imposing a certain normativity.

—MEXICAN SOCIAL DEMOGRAPHER

When I became department chair, I searched for articles on informality in mainstream urban studies journals. It turned out that articles on the impact of informality was about three percent when, in my opinion, that is *the* topic of the Latin American city because half of the city is informal! It was a sign that there are certain topics on which these journals are not interested at all.

—PERUVIAN URBAN SCHOLAR

In any field, at any time, when knowledge is highly centered and the infrastructures of production of that knowledge are strongly centralized, what happens is a phenomenon that I call epistemic indifference. In that case, on the one hand, the subjects excluded from that center are not recognized as cognitive subjects. This lack of recognition is neither new nor exclusive to groups considered marginal, since there have always been forms of knowledge that emerged from the experience of oppression and silence, such as exile or religious wars (Broncano 2020). Stories of women in science, the humanities, or other fields who made contributions that were utterly ignored in the mainstream narrative of the history of those fields abound (Harding 2016). Epistemic indifference is a form of production of nonexistence. It is more than mere ignorance or dismissal of an intellectual stance. It can come to constitute a form of epistemicide (de Sousa Santos 2018).

Seen from the center, epistemic indifference is a practice of decoupling. While ideas can flow from the center to the periphery in a kind of linear diffusionism, the center dismisses as irrelevant what emerges from those peripheries. The exception here is what emerges from the datafication of the periphery, that is, its transformation into data to sustain abstract and theoretical approaches produced in the center (Burke 2000; Couldry and Mejías 2019; Jöns 2011). In this process, actors of the center construct in more or less sophisticated ways reasons to justify their dominant position, ranging from issues of language ("they do not write well in English") to questioning the ability to argue ("this is literature, not a scientific article"). Dominance, however, is more the product of exclusionary practices, such as conferences in a single language, strict guidelines for academic writing, or canonical forms of data presentation, than the result of a quality differential (Mora et al. 2020; Neubert 2022). How could centers' scholars know such a differential if, at bottom, they ignore what is produced outside in the elsewhere?

From the point of view of the periphery, epistemic indifference plays three roles at once. First, it is a criterion of academic hierarchization. In general, the center ignores the periphery. But when someone or something from the periphery is recognized in the center, then that person or that academic product is assumed to be valuable. Whoever transcends from the periphery is almost automatically considered a success story, a model to imitate, a fully fledged cognitive subject, an authority. There is no lack of cases in which a rather personal link between an academic from the center and one from the periphery ends up translating into the latter becoming an authority on what happens in their sphere of action. And yet the center academic lacks a profound knowledge of the other's ideas.

Second, epistemic indifference places on the actors of the periphery the burden of proving their worth, whether by learning the academic lingua franca, traveling to academic conferences, or publishing in central journals. Whoever succeeds, of course, will receive the benefits of that intellectual blessing. Those who do not—and here lies an important ethical dimension—are condemned to irrelevance for being parochial, interested in local issues, and unable to master the rules of the global academic game (Sinha-Kerkhoff and Alatas 2010). Interestingly, at such a stage or in such a context, it is almost impossible for the periphery person to break the barrier of epistemic indifference. Perhaps one of the few ways of breaking it has been academic mobility of students and academics toward the centers, sometimes due to exile or reasons of economic-social crisis (Ude 2022). These subjects, once they get credentials and are back in the periphery, have reclaimed a certain authority that had been granted in the centers, with its corresponding rituals. An example is obtaining a doctorate in an institution of the center.

In the third place, epistemic indifference means that actors in the periphery are confronted with research agendas that rarely relate to the urgencies and needs of that periphery. Paradoxically, institutions in the peripheries demand from their academics the symbolic recognition of participation in the central academy, such as publications in high-impact journals. To do so, however, academics sacrifice pertinence and relevance at the local level (Alatas 2001; Neves 2022).

Epistemic Extractivism

> [In the context of this agreement with universities in the North] normally between three and six master's students come here to do fieldwork, because they do studies on Mexico. So, they know how to gather information, they get all the information from here and take it with them, but I wouldn't call that internationalization.
>
> —CHILEAN GEOGRAPHER

The inclusion of ideas previously considered marginal in some debates of central institutions leads to discovering what is produced elsewhere, by other actors, in other contexts. Intellectual novelty appears as an indicator not of factual novelty that already existed, but of previous ignorance that was not known. A race is unleashed, then, to nourish the local debate with ideas, perspectives, points of view, theories, and knowledge produced elsewhere, in other ways, by other cognizant subjects or hidden in distant infrastructures such as archives that have not been previously studied. But, in this case, the infrastructures are not developed for such knowledge to arrive directly from those margins. Instead, they must be translated by spokespersons from the center who become gatekeepers of that previously ignored or undervalued knowledge. Certain debates between US feminists and their counterparts in Africa or Latin America have shown this (Rodó-Zárate 2021). Similarly, disagreements between some European post-colonialists and theorists and researchers from former colonies, such as India or Mexico, point in the same direction (Grosfoguel 2022; Patel 2021; Restrepo 2016). The question is, as asked by postcolonial thinkers, who can or should speak for the other? And that question points to the reasons why decentering can never be considered complete until epistemic decentralizing is advanced and consolidated. Put differently, someone will always speak for the other until the other has a way to speak for himself or herself—beyond his or her local environment.

The reasons why some pieces of knowledge produced at the margins are of interest in debates in the centers is because they have some form of value, just as traditional extractivism (of natural resources, for example) is premised on the idea that such a resource can be capitalized in trade relations. Following what Alcoff discusses in detail in chapter 1, we can say that knowledge is commodified when (1) the value of knowledge is considered universal—and therefore its context of production no longer matters; (2) when the value is not relational—that is, it does not depend on continuing to be contained in the epistemic communities that gave rise to it; and (3) it can be protected through a legal and economic structure (e.g., with property or author rights). In such cases, as Alcoff (chapter 1) rightly argues, "knowledge must be reduced to information or data without involving interpretation or analysis, thus ensuring an inevitable epistemic injustice" and its consequent depoliticization by decontextualization. Epistemic extractivism has become an important ethical and practical problem in recent years. In the vocabulary proposed here, it consists of a set of practices where knowledge is decentered (or is being decentered) while infrastructures and their modes of operation remain centralized (or, worse, centralizing). Focusing on practices does not deny, however, that they require certain epistemological ideas that justify, as noted in chapter 1, noncooperation between actors. In any case, it delves into the materiality and practicality that makes possible the subordinate inclusion of objects and subjects from the periphery to nurture the journals and conferences, the classes and research projects, of the center. This is a familiar feature of metropoles largely independent of the good intentions of metropolitan academics (Simpson 2017).

The subordinate role that participants from the periphery have in the integration of international research teams, often led by Europeans or North Americans, has been studied (Kreimer 2006). It has been repeatedly argued that the agenda and theoretical frames of reference are set from positions of power in such groups. These are often associated with national and supranational funding structures (Losego and Arvanitis 2008). The situation is similar for objects, artifacts, and technologies. The tendency of northern institutions, for example, to acquire, legally or illegally, objects obtained in fieldwork and exploration in the peripheries is long-standing and has given rise to debates about restitution (Hicks 2020). Museums, zoos, botanical gardens, and archives that were once the epistemic crown jewels of overseas empires are today under severe criticism for their refusal to acknowledge the usurping role of metropolises and the illegitimacy of keeping other peoples' collections and objects of cultural value under the control of a few northern states (Hauser-Schäublin and Prott 2016). To hoard that which was not validly obtained is a way of producing helpless, incomplete, partial cognitive subjects. From there to ignoring their intellectual production or demanding their accreditation in the centers of knowledge production to have a valid voice is only one step.

The situation is no different if the acquisition has been legal, as illustrated by the purchase of the correspondence of Latin American writers by American universities. Epistemic extractivism is not possible if the infrastructure does not tend to centralize, to concentrate in certain places or hands. Because, in the end, with the concentration of information, the sociotechnical network that allows the production of new and relevant knowledge around it is justified and makes sense.[1] This was very accurately recognized when the University of Texas acquired part of Gabriel García Márquez's correspondence after his death:

> More than 150 UT faculty members engage with Latin America inside and outside the classroom walls, and numerous graduate and undergraduate students choose our university for its emphasis on Latin America. With outstanding patience and perseverance and care, for almost one hundred years the Nettie Lee Benson Latin American Collection has assembled one of the premier libraries devoted to the region. The Teresa Lozano Long Institute of Latin American Studies (LLILAS) integrates more than 30 academic departments across the university and ranks among the world's leading centers of its kind. All this to say that, *in its new home at The University of Texas at Austin, the Gabriel García Márquez Archive is surrounded by a rich cultural milieu and finds itself in fertile ground for intellectual discussion.* So much for remorse, let the celebration begin. (Montelongo 2015, 6, emphasis added)

It is not surprising then that, shortly afterward, it was announced, "From October 28–30 (2015), a three-day symposium, *planned by Latin American Studies faculty at the University of Texas at Austin,* will celebrate the life and work of Gabriel García Márquez. Interested people around the world will be able to watch and listen to the symposium panelists via a live webcast where *scholars will guide us toward a better understanding of García Márquez's life and accomplishments*" (Enniss 2015, emphasis added).

Epistemic centralization, which requires substantial resources, is always an inconclusive process. Institutions in the center recognize that the possession of these resources justifies the investment largely because they understand the benefits of the "better understanding" that comes with it. The periphery, on the other hand, generally discursively recognizes the value of these resources but simultaneously underfunds the necessary infrastructure and corresponding practices.

Epistemic Co-optation

> When I studied at CIESAS [Centro de Investigación y Estudios Superiores en Antropología Social] Occidente, CW invited me to accompany him to do field research in the highlands of Jalisco, in Tlaquepaque, along with students from the University of California and six CIESAS students. I got involved in that ten years ago and from then on I began to accompany them on different occasions to other communities to do research. I accompanied them on four occasions and that opened the doors for me to publish, because the final product of the projects was a book published by the University of California.
>
> —MEXICAN ANTHROPOLOGIST

> At a congress in Spain I was introduced to Salvador and he said to me: "Hey! I'm going to apply for a project to the European Union with the University of Salamanca. The project has to do with Latin America and you are Latin American scholars." We had a project that we had presented to Conacyt and that had been approved, but they hadn't given us funds for a survey we wanted to do, which was 80% of the budget. I told him, "Well, look, they had just approved a project on subnational democracies but has been only partially funded." So he said, "Ah, perfect. I'll put together the team in Spain, you put it together here in Mexico, and we'll do the project."
>
> —MEXICAN POLITICAL SCIENTIST

In 2006, I received an invitation from the president of the institution to (participate in) a project in France on the globalization of the Mediterranean. It was a study of borders and I had been working on industrial relocation processes. At that time our European counterparts had more experience working on Morocco, so they asked us to choose several areas or a comparative area of borders that we could investigate. I see that the case of Spain-Morocco is very similar to that of Mexico-United States, because they are areas with different levels of development. I decided to go to (and focus on) Morocco.

—MEXICAN DEVELOPMENT SCHOLAR

When knowledge is centered, with canonized theories and methodologies influencing the length and breadth of research agendas, undergraduate and postgraduate teaching, and ways of thinking about the discipline and its problems, the form of knowledge production may also begin to decentralize. How? By gradually expanding infrastructures and practices through the incorporation of actors who are located at the margins (Essanhaji and van Reekum 2022; Fasenfest 2022).

The inclusion of scholars from developing countries on the editorial boards of mainstream journals or the digitization of archives originally located in the periphery by universities or libraries in the developed world are examples of academic co-optation (Invernizzi et al. 2022). At first, scholarly co-optation seems to satisfy both sides. While the center diversifies, often to meet participation quotas of minorities or vulnerable groups (Morgan et al. 2022; Mousa 2021), the periphery receives at least some of the symbolic capital that comes with involvement in these projects and initiatives. What flows to the center, as has been mentioned many times, is data or subordinate specialized work that, in general, seeks to corroborate existing knowledge in the field.

The participation of co-opted actors is guaranteed to the extent that they do not undermine the bases and foundations of that knowledge. Participation cannot be politicized, understanding politicization as a reflection and potential transformation of power relations. Thus, invitations are often extended to actors on the periphery who can be coupled to or domesticated within the dynamics, practices, and infrastructure of the center. This includes both people and objects. The scholar invited as an editorial board member to the journal generally has the credentials necessary for international recognition, such as graduate study at some metropolitan institution. And the objects integrated tend to operate as immutable mobiles

(Latour 1987; Stöckelová 2012) that can be articulated with those already located in the center. An example here is digitized collections that are integrated into servers at northern institutions.

Although epistemic co-optation does not necessarily impede diversity (it may bring more criticism and trigger potential decentering), it does not seek it, nor does it usually produce it. Differences in prestige and resources, among other gaps, tend to operate so that those who co-opt can impose an agenda, a theoretical perspective, or a research methodology. Or, at least, they can hinder the emergence of alternative positions. For those who join initiatives from the center with the possibility of being part of projects whose extension and impact would be difficult to match with resources from the periphery, the situation is perceived as highly advantageous. Such initiatives will swell a curriculum that may allow repositioning in the local field thanks to these international contacts (Rodriguez Medina et al. 2019). The price of this type of integration to international agendas and institutions seems low for those from the margins who insert themselves into this type of project. It would be unfair to assert that epistemic co-optation has no potential to empower previously marginalized actors. However, to the extent that actors on the periphery maintain low levels of influence over the dynamics of knowledge production, such empowerment tends to remain latent.

Epistemic co-optation can be oriented toward nonhumans, from animal or plant species indigenous to remote regions to archives, collections, and databases. The history of academic collaborations in which scholars from the periphery participate in teams because they have access to species indigenous to their regions is vast, rich, and sometimes illustrative of certain injustices (Hurtado 2012). Species appear co-opted, often through the actions of local academics or technical personnel, within vast networks of institutions that tend to centralize knowledge. The participation of these intermediaries is what distinguishes epistemic co-optation from epistemic extractivism, where it is the experts from the center who arrive directly. Here, these enlisted local actors are not simply intermediaries but actors with the capacity to produce effects. Interestingly, some of these effects can be seen as distortions from the eyes of the centers of these global networks, given that in the peripheries there may be different ways of cataloguing, processing, classifying, ordering, and organizing information (Mutula and Tsvakai 2002).

Even when some nonhumans, such as the fauna and flora, cannot be displaced except by being translated into drawings, diagrams, or graphs, there are others that, because of their capacity for articulation, are eventually brought to the centers. I have participated in at least two international investigations where the interviews I conducted in Mexico ended up archived at the National Science Foundation. The interest in data, in gathering the information by co-opting the transcriptions of interviews, was so great that the contract involved my team not only conducting the interview in Spanish and transcribing it, but also translating it into English (albeit using specialized software) and then sending the package of transcribed and translated interviews to the United States. My research colleagues and I were co-opted, but so was the data.

Epistemic Mutualism

I came with my own project, which was relatively obvious and simple. Harvard had been doing research in Chiapas for thirty years, had a project called the "Chiapas Harvard Project" that began in 1957. A Harvard researcher, an anthropologist came to Chiapas and settled down. He connected with the Mexican *indigenistas* and started bringing in graduate and undergraduate students to do their theses on the indigenous peoples in Zinacantán. He directed that project for almost thirty years, brought one hundred and forty students to Chiapas and then finished his career, kept coming back and bringing people until his death in 2004. His entire archive is stored at Harvard, nobody uses it. So, I arrived and said: "Let's put this material to work!" The project was rescued, I worked on that archive. The people in charge of the archive are archaeologists from the Harvard Museum. I met them, I told them what it was about, I told them that I was a researcher from Mexico who wanted to do an ethnography and I asked them to let me in. They were super nice. And this year, not surprisingly because they had already told me they were going to invite me back, they said, "This year we have [funding] for a week. So, I put [funding for] the other week and I stayed for fifteen days. The archive is so big that the time I was there doesn't give me time for much. It's like fifteen drawers, more boxes, thousands of things, without a catalogue, just raw material. Maybe I'm going to catalogue it, so that the university can later make it available to the public. I'm fascinated because they give me the chance to do many things. The colleagues were the ones who received my project. When I went there they involved me in [many things]: introducing so-and-so, taking me to such-and-such a place, they gave me things to read. It's very "cool," very interesting. The idea is that with them I can elaborate a project for the complete digitization of the archive and then in some way I can intervene in the digitization and perhaps get funding.

—MEXICAN ANTHROPOLOGIST

As fate would have it, I was invited to UNAM [Universidad Nacional Autónoma de México] and a colleague from Cornell, from the school of labor relations, was there. She asked me to take her to the hotel, we had a traffic jam for several hours and then it occurred to us that we had to hold a meeting at Cornell with specialists. "I call all the Mexicans here and you call all the Americans and Canadians." And she said, "Sensational!" In a little while we put together the theme, we put the authors and she went back to Cornell, got funding, which is not difficult in the United States, and we did the event before the signing of the North American Free Trade Agreement.

—ARGENTINE LABOR SCHOLAR

When knowledge is decentered (with, at least, multiple ideas, worldviews, or perspectives coexisting as alternatives) and the mode of production, its infrastructures, and its practices, is decentralized, then we are in a scenario of epistemic mutualism. Epistemic mutualism is the genuine recognition, through all the mechanisms legitimized by the center and materialized in practices and infrastructures, of the knowledge produced by peripheralized actors.[2] This is observed when previously marginalized actors, even nonhumans (Pickering 2005), are treated as equals with respect to actors located in the centers of knowledge production. Such treatment does not imply the need to perceive themselves as equals in the strictest sense of the term; that is, it does not require losing the specificities that come with the situationality of each individual or group. Being in the periphery or in the center entails differences that, rather than being hidden, should be uncovered, highlighted, and, as far as possible, recognized (Kunce 2022).

Given the context of widespread asymmetries in global science, epistemic mutualism is a fragile and precarious situation that cannot be taken for granted unless there are ways of producing, maintaining, and extending it. For this reason, it is frequently observed almost anecdotally, in small stories that slip into the interstices of the great narratives on the unequal production and circulation of knowledge. In this sense, the dualism of epistemic asymmetry can be overcome, partially and momentarily, if we can imagine fractionalities (Law 2002, 4) in which epistemic authority is effectively distributed from the center to the periphery. This is because, as Law indicates, "we need to understand that to make a center is to generate and to be generated by a noncenter, a distribution of the conditions of possibility that is both present and not present" (2002, 112–13).

Often, perhaps through lack of knowledge, ethnocentrism, or overconfidence in their own practices and infrastructures, actors in the center tend to be overvalued, and they undervalue their peripheral counterparts. Simultaneously, due to the global reach of hierarchies, actors in the periphery tend to overvalue their counterparts in the center and undervalue themselves. These mechanisms are some of the ones that permanently create and re-create centers and peripheries of knowledge production (Rodriguez Medina and Vessuri 2021). It should be noted, however, that given that centers and peripheries are not homogeneous spaces, in both there are actors who can support initiatives that seek centralization or decentralization. In both social spaces there are, then, actors who reinforce epistemic injustices and actors who seek to counteract them. As Medina argues, "the particular obligation with respect to hermeneutical justice that differently situated subjects and groups have are interactive and need to be determined relationally. Whether individuals and groups live up to their hermeneutical responsibilities has to be assessed by taking into account the forms of mutual positionality, relationality, and responsiveness (or lack thereof) that these subjects and groups display with respect to one another" (2013, 90). It is under this assumption that there are precarious arrangements that allow the existence of peer relationships even in contexts of asymmetry, arrangements that the processes of epistemic decentralization should identify and reinforce, until eventually institutionalizing fairer epistemic orders.

In order for epistemic authority to be distributed, it is necessary to establish practices and infrastructures that, like valves, stimulate flows of circulation or of material and symbolic resources toward the margins, not in an occasional and casual manner, but in a permanent and intentional way. Thus, these flows would be key to producing new spaces of peerness. To such an extent are these spaces crucial that Turnbull has pointed out that the future of local knowledge traditions depends on the creation of "third spaces." These are interstitial places where these traditions can be reframed and decentered and where the social organization of trust can be negotiated (Turnbull 1997, 560). Actors on the periphery cannot be seen as passive recipients of such authority but only as interlocutors from whom learning should and can take place. As Connell (2007) argues, centers lose much by ignoring what happens beyond their comfort zones.

Why would equal treatment be desirable? As numerous studies in the field of science, technology, and society (STS) have shown, stakeholder diversity tends to produce better knowledge (Harding 2015). Studies abound pointing out that there are discoveries that would not have been possible if not for the inclusion of perspectives different from the mainstream (Nielsen 2011), those that have shown that there are alternative paths of science and technology development that were not explored (Frickel et al. 2010; Hess 2009), and those indicating that peer review could be improved with the implementation of decentralized systems focused on reviews and not on reviewers (Tenorio-Fornés et al. 2021). In the field of the internet, it is accepted that the centralization that large corporations have accentuated lends itself to abuse and that "what is necessary is for the generalized skillset of scientific, logical and algorithmic thinking that underlies programming to be spread throughout the population . . . in order to maintain autonomy in the era of the Internet" (Halpin and Monnin 2016).

In the same vein, what is called citizen science has shown that the participation of marginalized actors in knowledge production processes brings to light organizational forms of data collection or agenda setting that lead scientific practice toward better results (Krick and Meriluoto 2022; Strasser et al. 2019). At the level of technological development, equal treatment is key to processes of decentralization in the production of knowledge-intensive goods, as COVID-19 vaccines have recently shown (Kollmann 2021; Goodman et al. 2021), even when the postcolonial undercurrent is still clear in the rejection from the center of pandemic crisis management in the peripheries (Meghji and Niang 2022). Also at the governmental and intergovernmental levels, literature has shown that people's knowledge is often common sense and tacit, so their perspectives and values may be unavailable to others, especially decision-makers (Niskanen 2001; Ward 2021).

What cannot be ignored is that with diversity comes epistemic problems: How to reconcile different worldviews or ontologies between, say, indigenous groups and pharmaceutical companies hoping to exploit their ancestral medicinal and herbal knowledge (de la Cadena and Blaser 2018)? How to find alternative means of communication of knowledge to the hegemonic ones (patents, articles, etc.) and thus respond to the needs and interests of broader groups than limited professional communities, such as the scientific one (Albornoz et al. 2020)? How can we create controls to prevent the circulation of fake information and knowledge while accepting that the very idea of falsehood or truth is at stake in the process of knowledge production and circulation (Mills et al. 2021)? Can we build databases in the metropolises that contribute to a deeper understanding of their own colonial past and that can be useful in revising it (Avignolo 2021)?

It would be utopian to be able to answer these questions with a single, infallible recipe, but I would venture to assert that for epistemic diversity to occur, authority to flow to the margins, and knowledge production to decentralize, actions to transform the current spaces of knowledge production will be necessary. Such spaces, as Okune et al. and Kuo point out (chapters 6 and 8, this volume), are possible to construct to the extent that infrastructures, practices, and knowledge are articulated with a goal that, for lack of a better word, we could call emancipatory or, as Masaka (2021) calls it, epistemically liberatory. These spaces are always liminal insofar as the humans and nonhumans who occupy them are undergoing a transition from a before to an after, from a time of hierarchies to a time of community building. They are spaces of fearlessness and generosity, as Lobo (2022) has called them, in which systemic vulnerabilities and inequalities can be negotiated. In epistemological terms, what characterizes these spaces is the presence of different epistemic norms. Following what we learned in chapter 1, we could affirm that epistemic decentralizing is possible if we recognize that all knowledge is incomplete, based on multiple ontologies and epistemologies, and that its production and circulation demand epistemic humility—from all parties involved—so that the encounter between these epistemic communities is, in the end, the true value to be recognized in each cognitive project.

Epistemic Decentralizing: Two Latin American Cases

The first case of epistemic decentralizing, closer to the social sciences, involves the restitution of the archaeological remains found in the early twentieth century in Machu Picchu by Yale professor Hiram Bingham.[3] Like any history of decentralizing, this one also begins with one of centralization and canonization. According to Ricardo D. Salvatore,

> After the discovery of Machu Picchu in 1911, Professor Hiram Bingham of Yale College organized a far-reaching program of exploration and research in the southern Peruvian Andes. The Yale Peruvian Expedition lasted, through various stages, from 1911 to 1916. After setting up camp at

Ollantaytambo, the expedition's members tried to collect, in addition to archaeological remains, all kinds of evidence about the region's topography, flora, fauna, minerals, indigenous lifestyles, and climate. In all respects, the expedition was a scientific milestone, enhancing the knowledge of Inca archaeology and producing dozens of scientific reports and papers in a variety of disciplines. (2003, 69)

Bingham wanted to take the objects to New Haven, where this evidence could be analyzed by experts from various scientific institutions, such as Harvard, the Smithsonian, or the Museum of Natural History. The debate between those who wanted to keep them and those who wanted to take them gained importance at that time and included nothing less than discussions about the location of science, since local archaeologists and indigenists argued that the natural location of the pieces was Cuzco and Lima (Salvatore 2003). Bingham's victory must be seen in the light of the competition between American and European archaeology, in which the United States had to demonstrate its superiority over its European counterpart in the same contested field: South America.

There was also a rivalry between Harvard and Yale as to which school could more quickly extract and better understand the secrets of Andean antiquity. As Salvatore argues, "if North American scientists could find and prove the true origins of Inca civilization (correcting the errors of Spanish chroniclers), South Americans would start to recognize the potential of the colossus of the North and question their blind admiration for Europe" (2003, 72). What was at stake, once the evidence was centralized, was the possibility of constructing a new and alternative hegemonic interpretation, the canon, of one of the most important civilizations of antiquity: the Incas. In the dispute between Europe and the United States, one of the battlegrounds was archaeological publications. Thus, centralization precedes centering, but it is the construction of the mainstream that is ultimately at stake.

After numerous diplomatic negotiations, in 2010 an agreement was signed for Yale to return what Professor Bingham had taken. The centenary of the discovery, in 2011, was used to claim the restitution, which finally materialized in 2012. On November 12, 2012, the last shipment arrived: 35,000 pieces in 127 boxes (Fowks 2012). This completed the transfer of more than 46,000 pieces in three deliveries between 2010 and 2012, of which more than 4,800 were declared National Cultural Heritage in 2019 by the Peruvian government (Andina 2019).

This claim and restitution of historically valuable pieces is not unique. It is part of a wider movement that has some of the world's most famous museums, such as the British Museum, at the center of the storm (Hicks 2020). The central idea that dominates current debates about the restitution or return of cultural property is that an artifact, especially if it is considered sacred or crucial in the generation of identity, belongs to its community of origin (Hauser-Schäublin and Prott 2016, 2). What is sometimes lost sight of is that, in addition to its symbolic identity value, restitution implies the possibility for marginalized actors, through the reappropriation of their objects, to produce alternative knowledge to that accepted as true and frequently exclusively validated by metropolitan academia. Moreover, as Alcoff (chapter 1, this volume) points out, the disputes between museums and epistemic communities that have had part of their material culture taken away from them is also a "struggle . . . over ontological assumptions about how to conceptualize objects as well as how to prioritize diverse conceptions of value." While the focus on ontology can show the different "equivocations" that are inherently implicit in all intercultural translation (de la Cadena 2015), the rest of the case we address here attempts to show the capacity to produce a certain dialogue—at least at the academic level—that shows the value of the epistemological dimension of restitution.

In this context, it was not surprising that the findings of Professor Burger of Yale University and his team on the earliest settlement at Machu Picchu, which appeared in the prestigious journal *Antiquity* in 2021, were published after those of another team, which included archaeologists from Peru charged with safeguarding the material returned by Yale. That team arrived at similar results, disseminated them at a conference in Arizona in 2019, and published them in the journal *Radiocarbon* in 2021 (Ziółkowski et al. 2021; Burger et al. 2021).

What was now known was that Machu Picchu already had been populated two decades earlier than previously thought, around 1420, defying interpretations based on Inca texts. One of the Peruvian archaeologists and coauthor of the study noted that "Burger's team measured samples from burial caves, 'from individuals who were recovered by the expedition of Hiram Bingham,' the American credited with the discovery of the Inca citadel between 1911 and 1912. They measured the remains before returning the archaeological and skeletal materials to Peru (Fowks 2021). It goes in part against the logic of centers and peripheries that Burger's research only confirms the pioneering research conducted with the participation of Peruvian scientists, who note, with some derision, that "with respect to the dating of North American colleagues, we are happy that their results are quite similar to ours" (Fowks 2021).

What is at stake is not a race between research teams but the possibility for scientists on the margins of world science to access the material with which the past can be rethought. Or, in other words, at stake is the construction of another infrastructure, with which to produce cutting-edge knowledge about their own society or past. As Laely points out, restitution is "a fundamental endeavor to diversify knowledge creation and bring in multiple voices—not attempting to produce a 'shared heritage,' but to share knowledge on the basis of, amongst other things, a (at times) shared history" (2020, 30).[4]

The second case of epistemic decentralizing involves the development of the most complex and expensive infrastructure we can imagine, the astronomical one, and it is located in the Atacama Desert, in the north of Chile. In a newspaper article, it was proudly pointed out,

> "It is very possible that the next Chilean Nobel Prize winner will be an astronomer," says Mario Hamuy, director of the Department of Astronomy at the University of Chile. The figures on the productivity of the discipline in the country seem to prove him right. Today, according to the Scientific Information Program of the National Commission for Scientific and Technological Research (Conicyt), astronomy is the most productive discipline in the country. Between 2003 and 2010 it has registered 3,147 international publications, 11% of national productivity. Mónica Rubio, director of Conicyt's Astronomy Program, says that between 150 and 200 research papers are published annually by Chilean authors, the scientific area with the highest level of production per scientist. "Chilean astronomers working in Chilean institutions are more productive than the world average. On average, they have three publications per year, above the international average, which is a little less than 2 (1.8–1.7)," she says. "Astronomers in more developed countries that have 90% (use of telescopes) don't seem to be as efficient. Chilean astronomy is very prestigious abroad," says Rubio. (Espinoza 2010)

How is it possible for a peripheral country like Chile to produce science of such caliber (and relative quantity) as to aspire even to a Nobel Prize? The story of this explosion of Chilean astronomy is undoubtedly connected to the construction of the Atacama Large Millimeter/submillimeter Array (ALMA) in 2013, a clear example of epistemic decentralizing. Composed of fifty-five radio antennas, ALMA is under the control of a consortium with representatives from North America, Europe, and East Asia (Lehuedé 2021, 20). The observatory was located in northern Chile because it is considered one of the best in the world for astronomy due to the absence of clouds most of the year (about 340 days of clear skies), dry soils, and low light intensity or pollution. Already by 2018, 40 percent of all astronomical observations worldwide came from this region (Cortes, Depoortere, and Malaver 2018).

It cannot be overlooked that the Chilean state strongly contributed to the arrival of these ventures through an official discourse on infrastructure as development (Lehuedé 2023) and even with tax benefits for the observatories (Barandiaran 2015). The implementation of the observatory had a clear impact on academic production. Between 2005 and 2015, the annual productivity of Chilean papers tripled. The year 2014 stands out, with a 30 percent increase in productivity compared to previous years. Also, the Pontificia Universidad Católica de Chile, the Universidad de Chile, and the Universidad de Concepción became the most productive institutions, even compared to counterparts in the Global North (Cortes, Depoortere, and Malaver 2018, 3).

While ALMA could be considered a case of epistemic co-optation, where Chilean resources are placed at the service of agendas and projects designed and directed from global academic production centers (Guerrero and Arroyo 2014, 212), the situation is actually more complex. On the one hand, the scientific community has questioned some scientific policies that include the evaluation of astronomical projects, the massive sending of students to train abroad, and the valuation of academic productivity only on the basis of indexed articles (Barandiaran 2015, 161).

On the other hand, the increase in the quantity and quality of Chilean publications in astronomy was produced through an agreement where the local astronomical community has access to 10 percent of the observing time (Lehuedé 2021), which has been seen as sufficient given the size of the community. However, the construction of a new observatory, the Vera C. Rubin Observatory (formerly called the Large Synoptic Survey Telescope or LSST) now provides the opportunity for more equal recognition in the international academic community. Lehuedé notes the case of a team of astronomers from the city of La Serena who, when asked about international links, state,

If you approach [the observatories] as an educational entity with first-level scientific collaboration teams, the communication is much better. It is one of peers. The observatories only did outreach here for a long time. And now that we have a doctoral program and an astronomy group . . . now we are partners. We can offer them an alternative route for their networks . . . and that is not *gringo* or anything. So, when you establish a relationship in terms of "I give to you and you give to me," their attitude is quite good. (2023, 432–33)

Epistemic decentralizing leads peripheral actors to think differently about their global academic relations, even if the asymmetries do not disappear. There are benefits for both parties. They consider each other as partners or peers and aspire to more egalitarian positions, which has been made possible by the level of scientific research and technological infrastructure achieved (Lehuedé 2023, 432–44). The reference to doctoral programs and multinational research teams is not a mere illustration in the Chilean astronomer's statements, but the confirmation that the materialization of the link is a necessary condition for the construction of peerness.

The story would not be complete if we did not see that the inclusion of Chilean astronomers is only the first step in a process of decentralization that is always dynamic. Lehuedé points out that "the implementation of data-intensive research globally [leads to] datafication that might be acting as a rationality that some communities are rushing to adopt without asking fundamental questions regarding *with whom* and *for whom*" (2023, 438). Logically, one possible answer to those questions is with Chilean scientists and for Chilean science. Another response, on the other hand, would seek to broaden the spectrum of actors affected by large infrastructure projects and take into account, for example, the ontologies and epistemologies of the communities that inhabit these regions, such as the Lickan Antay people in northern Chile. "Just like large-scale mining and monocultures, an expansion of data infrastructure predicated upon an assetized ontology involves the reduction of a pluriverse, where different worlds coexist, to a universe, where only a single world of capitalist and modern contours is allowed to thrive" (Lehuedé 2022, 15).

Can the process of epistemic decentralizing ever be completed? To the extent that there are always practices that can be more horizontal, infrastructures that can allow for more pluralism, and individuals and groups that can be enlisted in wider and denser networks of knowledge production (and circulation), the process of decentralizing will always be open.

There may be, as we showed in the previous section, some developments that leave open the possibility of further increasing the number and variety of active participants. There may also be setbacks, as epistemic centralization is also an open and largely successful process (Schneider 2019).

Conclusion

In this chapter we have followed a path that began by pointing out the processes through which we can characterize knowledge and its conditions of production. We have presented four scenarios: (1) epistemic indifference, (2) epistemic co-optation, (3) epistemic extractivism, and (4) epistemic mutualism. These scenarios emerge from crossing two variables that, often, especially in STS, have been perhaps excessively kept together. On the one hand, there is the decentering of canons and mainstream currents that have homogenized what is really heterogeneous and have produced and maintained hierarchies between epistemic subjects. Some thoughts, however local they may be, appear global, or so they are presented, while others, however universalistic their scope or aspirations, remain local (Mignolo 2000). On the other hand, decentralizing infrastructures and practices is based on the recognition that the inclusion of the other is not only a matter of identifying and recovering their ideas but of transforming the conditions of production for their integral participation as peers. Regardless how we understand epistemic justice, it has to do with ideas and infrastructures at the same time.

In the second part we provided two examples, brought from the scientific reality of Latin America. Chilean astronomy and Peruvian archaeology illustrate processes of epistemic decentralizing in action. In both, the focus is on materiality and infrastructure. Peruvian archaeologists recover objects gathered in expeditions carried out by US colleagues in the early twentieth century. Chilean astronomers take advantage of the construction of state-of-the-art observatories in the Atacama Desert. In both cases, such infrastructures or materialities translate into new knowledge, produced from the periphery. Peruvian archaeologists challenge accepted knowledge about the origins of the population of Cuzco, while Chilean astronomers exponentially multiply their articles, making them the most productive within the international astronomical community. Finally, the limits of epistemic decentralizing come to light. In the Peruvian case, it is pointed out that the publication of the findings by the Yale team validates what was found by the team involving Peruvian academics, which had previously presented the evidence for the settlement of Cuzco. The validity of the knowledge has yet to be decentralized (Fowks 2021). In the case of Chile, the construction of observatories in territories inhabited by indigenous communities raises controversies about the real possibility of the knowledge produced being put at the service of local problems (Lehuedé 2022). The lack of recognition of groups as cognitive subjects indicates unfinished epistemic decentralizing.

The decentralization approach may seem a step backward for those who, in STS, had sought to remove the barriers between the cognitive and the material, between thought and technology, between ideas and infrastructure. A first reading of this chapter might indicate that these dichotomies are being reinstated. However, this is not the case. What we seek here is to show that the elements that make up heterogeneous assemblages of knowledge production can have diverse characteristics and be subject to different processes. The very heterogeneity of the assemblages (Law 1992) indicates that there is diversity and, with it, opens the possibility of differentiated dynamics. To recognize, for example, that only individuals can be moved by passions or that technologies are the ones that can have functions, strictly speaking, is to point out something proper to each type of element. It is to do so without denying that, in order to achieve certain effects, such as producing knowledge, they must act in entanglements. Nor can it be denied that such entanglements have in turn some emergent properties that their constituent elements do not possess; they are "network effects" (Law 1999).

The nodal point of the idea of epistemic decentralizing is to point out that there is no possibility of empowering actors, human and nonhuman, in the periphery without a simultaneous change in the disciplinary mainstream and in the infrastructures and practices of knowledge production. Empowerment does not end with denouncing in highly centralized infrastructures the oppression of certain groups and their epistemic views. Empowerment does not happen either if we only open more spaces (infrastructures and practices) for the multiplication of actors while reproducing excluding ways of thinking.

Contrary to what some theories suppose, strongly decentered and decentralized disciplines or fields are characterized by a plurality of voices, even by a contradictory set of coexisting points of view. Put differently, in a scenario of epistemic decentralizing, actors can produce and disseminate any idea they wish. This is shown by the proliferation of antivaccine groups that rely on some form of expert knowledge (Gobo and Sena 2022), the use of digital technologies by drug mafias (Ake-Kob 2020), and the "rebel expertise" of subversive groups that allows the development of improvised antipersonnel mines (Pardo Pedraza 2020). To assume that only counter-hegemonic ideas can emerge from the peripheries is to underestimate the cognitive subjects that live there, the infrastructures that function there, and the practices that develop there (see Keim 2011).

Assemblages are always complex and can never determine a course of action without room for contingency. In that sense, and as the history of democracy teaches, the expansion of rights and the empowerment of groups that were promoted by progressive parties do not always translate into electoral victories. Rather, those same empowered groups often vote for parties that advocate the opposite ideas to those that made their empowerment possible. The same is true of empowered cognitive subjects and infrastructures (Kornberger et al. 2019). Their agency allows them a freedom of thought and action that constitutes their autonomy, that is, their foundation as active participants in assemblages of knowledge.

Notes

1   The critique outlined here does not ignore the demanding and funda-mental technical work of document maintenance, which is undoubtedly one of the strengths of global centers of academic production. In relation to the archive of García Márquez at the University of Texas at Austin, for example, such preservation work, which includes cataloguing, digitization, and conservation processes, is remarkable (Díaz Canas 2015).

2   For some, the opposite of extractivism is deep reciprocity (Simpson 2014), which implies "fair exchange in relations between human beings and in relations between humans and non-humans. . . . Extracting with-out giving back is the principle of destruction of life. To extract taking care to reproduce life and to give back what is extracted is an entirely different cosmological principle" (Grosfoguel 2022, 263). My preference for mutualism is that the word emphasizes not only actions (which may or should be relatively reciprocal) but also shared effects—the idea of consequences for both parties. Mutualism has also been used in ecologi-cal thinking to refer to multispecies benefits (Holland and Bronstein 2008), emphasizing that the exchange need not necessarily be reciprocal but beneficial to both parties.

3   Another interesting case of epistemic decentralization in Latin Amer-ica is that of the transnational networks of academics built in Colombia

(Tejada 2012) and the journal *Tapuya: Latin American Science, Technology and Society* (Rodriguez Medina 2024).

4   In his analysis of museums in Switzerland and Uganda, Laely mentions that "the primary goal and guiding principle was, and remains, *mutual* scientific and practical exchange. This is associated with developing a knowledge partnership and training for all the museums and researchers involved and it includes the *joint* debate of museological best practice, which has been implemented directly in the *exhibitions devised, prepared and curated jointly* throughout all phases, starting with their conception. *Setting the agenda together* in this way certainly proved to be essential. In the case under study, it was clearly agreed from the very start that the three partner museums must all be *equally* involved, having equal say and rights in all stages of implementation, including conceptualization" (2020, 30, emphasis added).

References

Ake-Kob, Aline. 2020. "Goffman and the Mafia: Shaping YouTube's Technological Affordances in the War on Drugs." *Tapuya: Latin American Science, Technology and Society* 3: 493–511.

Alatas, Syed Farid. 2001. "The Study of the Social Sciences in Developing Societies: Towards an Adequate Conceptualization of Relevance." *Current Sociology* 49 (2): 1–19.

Albornoz, Denisse, Angela Okune, and Leslie Chan. 2020. "Can Open Scholarly Practices Redress Epistemic Injustice?" In *Reassembling Scholarly Communications: Histories, Infrastructures, and Global Politics of Open Access*, edited by Martin Paul Eve and Jonathan Gray, 65–79. Cambridge, MA: MIT Press.

Andina (Agencia Peruana de Noticias). 2019. "Machu Picchu: 4,849 piezas repatriadas de Universidad de Yale ya son patrimonio cultural." *Agencia Peruana de Noticias*, November 10, 2019. https://andina.pe/agencia /noticia-machu-picchu-4849-piezas-repatriadas-universidad-yale-ya-son -patrimonio-cultural-772533.aspx.

Avignolo, María Laura. 2021. "Esclavitud en Francia: Una base de datos expone a todos los dueños de esclavos y el dinero que recibieron." *Clarín*, May 18, 2021. https://www.clarin.com/mundo/esclavitud-francia-base -datos-expone-duenos-esclavos-dinero-recibieron_o_XbY-cUx8Q.html.

Barandiaran, Javiera. 2015. "Reaching for the Stars? Astronomy and Growth in Chile." *Minerva* 53 (2): 141–64.

Broncano, Fernando. 2020. *Conocimiento expropiado: Epistemología política en una democracia radical*. Madrid: Akal.

Burger, Richard L., Lucy C. Salazar, Jason Nesbitt, Eden Washburn, and Lars Fehren-Schmitz. 2021. "New AMS Dates for Machu Picchu: Results and Implications." *Antiquity* 95 (383): 1265–79.

Burke, Peter. 2000. *Social History of Knowledge: From Gutenberg to Diderot.* Cambridge: Polity.

Connell, Raewyn. 2007. *Southern Theory: Social Science and the Global Dynamics of Knowledge.* Cambridge: Polity.

Cortes, Rodrigo, Denise Depoortere, and Lucina Malaver. 2018. "Astronomy in Chile: Assessment of Scientific Productivity through a Bibliometric Analysis." *EPJ Web of Conferences* 186: 05002. https://doi.org/10.1051/epjconf/201818605002.

Couldry, Nick, and Ulises A. Mejías. 2019. *The Costs of Connection: How Data Is Colonizing Human Life and Appropiating It for Capitalism.* Stanford, CA: Stanford University Press.

Dalmiya, Vrinda. 2016. *Caring to Know: Comparative Care Ethics, Feminist Epistemology and the Mahabharata.* New York: Oxford University Press.

de la Cadena, Marisol. 2015. *Earth Beings: Ecologies of Practice across Andean Worlds.* Durham, NC: Duke University Press.

de la Cadena, Marisol, and Mario Blaser. 2018. *A World of Many Worlds.* Durham, NC: Duke University Press.

de Sousa Santos, Boaventura. 2018. *The End of the Cognitive Empire: The Coming of Age of the Epistemologies of the South.* Durham, NC: Duke University Press.

Díaz Canas, Diana. 2015. "Preservando la vida y legado de Gabriel García Márquez." *Ransom Center Magazine,* October 21, 2015. https://sites.utexas.edu/ransomcentermagazine/2015/10/21/preservando-la-vida-y-legado-de-gabriel-garcia-marquez/#more-15089.

Enniss, Stephen. 2015. "Como un personaje de una de sus novelas, Gabriel García Márquez ha llegado a un lugar sin tiempo." *Ransom Center Magazine,* October 21, 2015. https://sites.utexas.edu/ransomcentermagazine/2015/10/21/como-un-personaje-de-una-de-sus-novelas-gabriel-garcia-marquez-ha-llegado-a-un-lugar-sin-tiempo/.

Espinoza, Cristina. 2010. "Astrónomos chilenos, camino al estrellato." *La Nación,* September 22, 2010. https://web.archive.org/web/20150930181248/http://www.lanacion.cl/astronomos-chilenos-camino-al-estrellato/noticias/2010-09-21/212220.html.

Essanhaji, Zakia, and Rogier van Reekum. 2022. "Following Diversity through the University: On Knowing and Embodying a Problem." *Sociological Review* 70 (5): 882–900.

Fasenfest, David. 2022. "Spiking the Sociological Canon." *Critical Sociology* 48 (4–5): 549–52.

Fowks, Jacqueline. 2012. "La Universidad de Yale entrega a Perú el último lote de tesoros de Machu Picchu." *El País,* November 13, 2012. https://elpais.com/cultura/2012/11/13/actualidad/1352797501_821292.html.

Fowks, Jacqueline. 2021. "Machu Picchu se estableció dos décadas antes de lo pensado." *El País,* August 4, 2021. https://elpais.com/ciencia/2021-08-04/machu-picchu-se-establecio-dos-decadas-antes-de-lo-pensado.html.

Frickel, Scott, Sahra Gibbon, Jeff Howard, Joanna Kempner, Gwen Ottinger, and David J. Hess. 2010. "Undone Science: Charting Social Movement and Civil Society Challenges to Research Agenda Settings." *Science, Technology, and Human Values* 35 (4): 444–73.

Gobo, Giampietro, and Barbara Sena. 2022. "Questioning and Disputing Vaccination Policies: Scientists and Experts in the Italian Public Debate." *Bulletin of Science, Technology and Society* 42 (1–2): 25–38.

Goodman, Peter S., Apoorva Mandavilli, Rebecca Robbins, and Matina Stevis-Gridneff. 2021. "What Would It Take to Vaccinate the World against Covid?" *New York Times*, May 15, 2021. https://www.nytimes.com/2021/05/15/world/americas/covid-vaccine-patent-biden.html.

Grosfoguel, Ramón. 2022. *De la sociología de la descolonización al nuevo antimperialismo decolonial*. Mexico City: Akal.

Guerrero, Pablo, and Mary T. K. Arroyo. 2014. "Base Policy on Evidence." *Nature* 510: 212.

Halpin, Harry, and Alexandre Monnin. 2016. "The Decentralization of Knowledge: How Carnap and Heidegger Influenced the Web." *First Monday* 21 (12). https://doi.org/10.5210/fm.v21i12.7109.

Harding, Sandra. 2015. *Objectivity and Diversity: Another Logic of Scientific Research*. Chicago: University of Chicago Press.

Harding, Sandra. 2016. *Whose Science? Whose Knowledge? Thinking from Women's Lives*. Ithaca, NY: Cornell University Press.

Hauser-Schäublin, Brigitta, and Lyndel V. Prott. 2016. "Introduction: Changing Concepts of Ownership, Culture and Property." In *Cultural Property and Contested Ownership: The Trafficking of Artefacts and the Quest for Restitution*, edited by Brigitta Hauser-Schäublin and Lyndel V. Prott, 1–20. London: Routledge.

Hess, David. 2009. "The Potentials and Limitations of Civil Society Research: Getting Undone Science Done." *Sociological Inquiry* 79 (3): 306–27.

Hicks, Dan. 2020. *The Brutish Museums: The Benin Bronzes, Colonial Violence and Cultural Restitution*. London: Pluto.

Holland, Julian N., and Judith L. Bronstein. 2008. "Mutualism." In *Encyclopedia of Ecology*, 5 vols., edited by S. E. Jorgensen and B. D. Fath, 2485–91. Amsterdam: Elsevier.

Hurtado, Diego, ed. 2012. *La física y los físicos Argentinos: Historias para el presente*. Córdoba: Universidad Nacional de Córdoba and Asociación Física Argentina.

Invernizzi, Noela, Pablo Kreimer, Amylcar Davyt, and Leandro Rodriguez Medina. 2022. "STS between Centers and Peripheries: How Transnational Are Leading STS Journals?" *Engaging Science, Technology, and Society* 8 (3): 31–62.

Jöns, Heike. 2011. "Centre of Calculation." In *The SAGE Handbook of Geographical Knowledge*, edited by John Agnew and David N. Livingstone, 158–70. London: Sage.

Keim, Weibke. 2011. "Counterhegemonic Currents and Internationalization of Sociology: Theoretical Reflections and an Empirical Example." *International Sociology* 26 (1): 123–45.

Kollmann, Raúl. 2021. "La Argentina suma vacunas: Llegan las fabricadas en México." *Página 12*, May 25, 2021. https://www.pagina12.com.ar/343545-la-argentina-suma-vacunas-llegan-las-fabricadas-en-mexico.

Kornberger, Martin, Geoffrey C. Bowker, Julia Elyachar, Andrea Mennicken, Peter Miller, Joanne Randa Nucho, and Neil Pollock, eds. 2019. *Thinking Infrastructures: Research in the Sociology of Organizations*. Bingley, UK: Emerald.

Kreimer, Pablo. 2006. "¿Dependientes o integrados? La ciencia latinoamericana y la nueva división internacional del trabajo." *Nómadas* 24: 199–212.

Krick, Eva, and Taina Meriluoto. 2022. "The Advent of the Citizen Expert: Democratising or Pushing the Boundaries of Expertise?" *Current Sociology* 70 (7): 967–73.

Kunce, Aleksandra. 2022. "Why Should We Cultivate 'the Difference' in Everyday Practices of the University?" *Higher Education Quarterly* 76: 671–82.

Laely, Thomas. 2020. "Restitution and Beyond in Contemporary Museum Work: Re-imagining a Paradigm of Knowledge Production and Partnership." *Contemporary Journal of African Studies* 7 (1): 17–37.

Latour, Bruno. 1987. *Science in Action: How to Follow Scientists and Engineers through Society*. Cambridge, MA: Harvard University Press.

Law, John. 1992. "Notes on the Theory of the Actor-Network: Ordering, Strategy, and Heterogeneity." *Systems Practice* 5 (4): 379–93.

Law, John. 1999. "After ANT: Complexity, Naming and Topology." In *Actor Network Theory and After*, edited by John Law and John Hassard, 1–14. Oxford: Wiley.

Law, John. 2002. *Aircraft Stories: Decentering the Object in Technoscience*. Durham, NC: Duke University Press.

Lehuedé, Sebastián. 2021. "Governing Data in Modernity/Coloniality: Astronomy Data in the Atacama Desert and the Struggle for Collective Autonomy." PhD diss., London School of Economics and Political Science. http://etheses.lse.ac.uk/4321/.

Lehuedé, Sebastián. 2022. "Territories of Data: Ontological Divergences in the Growth of Data Infrastructure." *Tapuya: Latin American Science, Technology and Society* 5 (1). https://doi.org.10.1080/25729861.2022.2035936.

Lehuedé, Sebastián. 2023. "The Coloniality of Collaboration: Sources of Epistemic Obedience in Data Intensive Astronomy in Chile." *Information, Communication and Society* 26 (2): 425–40. https://doi.org.10.1080/1369118X.2021.1954229.

Lobo, Michele, 2022. "Breathing Spaces of Fearlessness and Generosity in the Anglophone/Western University." *Geographical Research* 60 (1): 126–37.

Losego, Phillipe, and Rigas Arvanitis. 2008. "La science dans les pays non-hégémoniques." *Revue d'Anthropologie des Connaissances* 2 (3): 343–50.

Masaka, Dennis. 2021. "Knowledge, Power, and the Search for Epistemic Liberation in Africa." *Social Epistemology* 35 (3): 258–69.

Medina, José. 2013. *The Epistemology of Resistance: Gender and Racial Oppression, Epistemic Injustice, and Resistant Imaginations*. New York: Oxford University Press.

Meghji, Ali, and Sophie Marie Niang. 2022. "Between Post-racial Ideology and Provincial Universalisms: Critical Race Theory, Decolonial Thought and COVID-19 in Britain." *Sociology* 56 (1): 131–47.

Mignolo, Walter. 2000. *Local Histories/Global Designs: Coloniality, Subaltern Knowledges, and Border Thinking*. Princeton, NJ: Princeton University Press.

Mills, David, Abigail Branford, Kelsey Inouye, Natasha Robinson, and Patricia Kingori. 2021. "'Fake' Journals and the Fragility of Authenticity: Citation Indexes, 'Predatory' Publishing, and the African Research Ecosystem." *Journal of African Cultural Studies* 33 (3): 276–96.

Montelongo, José. 2015. "García Márquez's *pentimenti*." *Portal* 10: 4–7. https://repositories.lib.utexas.edu/items/62e53223-45ef-4072-b2d2-9c05083e8320.

Mora, Raúl Alberto, Gerald Campano, Ebony Elizabeth Thomas, Amy Stornaiuolo, Bethany Monea, Ankhi Thakurta, and James Joshua Coleman. 2020. "Decentering and Decentralizing Literacy Studies: An Urgent Call for Our Field." *Research in the Teaching of English* 54 (4): 313–17.

Morgan, Allison C., Nicholas LaBerge, Daniel B. Larremore, Mirta Galesic, Jennie E. Brand, and Aaron Clauset. 2022. "Socioeconomic Roots of Academic Faculty." *Nature Human Behaviour* 6: 1625–33. https://doi.org/10.1038/s41562-022-01425-4.

Mousa, Mohamed. 2021. "Academia Is Racist: Barriers Women Faculty Face in Academic Public Contexts." *Higher Education Quarterly* 76 (4): 741–58.

Mutula, Stephen M., and Mashingaidze Tsvakai. 2002. "Historical Perspectives of Cataloguing and Classification in Africa." *Cataloging and Classification Quarterly* 35 (1–2): 61–77.

Neubert, Dieter. 2022. "Do Western Sociological Concepts Apply Globally? Toward a Global Sociology." *Sociology* 56 (5): 930–45.

Neves, Fabrício Monteiro. 2022. "Some Elements of the Regime of Management of Irrelevance in Science." *Tapuya: Latin American Science, Technology and Society* 5 (1): 2035951.

Nielsen, Michael. 2011. *Reinventing Discovery: The New Era of Networked Science*. Princeton, NJ: Princeton University Press.

Niskanen, William A. 2001. "Bringing Power to Knowledge: Choosing Policies to Use Decentralized Knowledge." In *Knowledge and Politics*, edited by Riccardo Viale, 107–18. Berlin: Springer.

Pardo Pedraza, Diana. 2020. "Artefacto Explosivo Improvisado: Landmines and Rebel Expertise in Colombian Warfare." *Tapuya: Latin American Science, Technology and Society* 3 (1): 472–92.

Patel, Sujata. 2021. "Sociology's Encounter with the Decolonial: The Problematique of Indigenous vs That of Coloniality, Extraversion and Colonial Modernity." *Current Sociology* 69 (3): 372–88.

Pickering, Andrew. 2005. "Decentering Sociology: Synthetic Dyes and Social Theory." *Perspectives on Science* 13 (3): 352–405.

Restrepo, Eduardo. 2016. "Descentrando a Europa: Aportes de la teoría postcolonial y el giro decolonial al conocimiento situado." *Revista Latina de Sociología* 6 (1): 60–71.

Rodó-Zárate, María. 2021. *Interseccionalidad: Desigualdades, lugares y emociones.* Barcelona: Bellaterra.

Rodriguez Medina, Leandro. 2024. "On Epistemic Decentralising: Infrastructuring Knowledge beyond Global North." *Globalisation, Societies and Education*, 1–11. https://doi.org/10.1080/14767724.2024.2307876.

Rodriguez Medina, Leandro, and Hebe Vessuri. 2021. "Personal Bonds in the Internationalization of the Social Sciences: A View from the Periphery." *International Sociology* 36 (3): 398–418.

Rodriguez Medina, Leandro, Hugo Pablo Ferpozzi, Juan Agustín Layna, Emiliano Martin Valdez, and Pablo Kreimer. 2019. "International Ties at Peripheral Sites: Co-producing Social Processes and Scientific Knowledge in Latin America." *Science as Culture* 28 (4): 562–88.

Salvatore, Ricardo D. 2003. "Local versus Imperial Knowledge: Reflections on Hiram Bingham and the Yale Peruvian Expedition." *Nepantla: Views from South* 4 (1): 67–80.

Schneider, Nathan. 2019. "Decentralization: An Incomplete Ambition." *Journal of Cultural Economy* 12 (4): 265–85.

Simpson, Leanne Betasamosake. 2014. "Land as Pedagogy: Nishnaabeg Intelligence and Rebellious Transformation." *Decolonization: Indigeneity, Education and Society* 3 (3): 1–25.

Simpson, Leanne Betasamosake. 2017. *As We Have Always Done.* Minneapolis: University of Minnesota Press.

Sinha-Kerkhoff, Kathinka, and Syed Farid Alatas, eds. 2010. *Academic Dependency in the Social Sciences: Structural Reality and Intellectual Challenges.* New Delhi: Manohar.

Stöckelová, Tereza. 2012. "Immutable Mobiles Derailed: STS, Geopolitics, and Research Assessment." *Science, Technology, and Human Values* 37 (2): 286–311.

Strasser, Bruno, Jérôme Baudry, Dana Mahr, Gabriela Sanchez, and Elise Tancoigne. 2019. "'Citizen Science'? Rethinking Science and Public Participation." *Science and Technology Studies* 32 (2): 52–76.

Tejada, Gabriela. 2012. "Mobility, Knowledge and Cooperation: Scientific Diasporas as Agents of Development." *Migration and Development* 10 (18): 59–92.

Tenorio-Fornés, Ámbar, Elena Pérez Tirador, Antonio A. Sanchez-Ruiz, and Samer Hassan. 2021. "Decentralizing Science: Towards an Interoperable

Open Peer Review Ecosystem Using Blockchain." *Information Processing and Management* 58 (6): 1–17.

Turnbull, David. 1997. "Reframing Science and Other Local Knowledge Traditions." *Futures* 29 (6): 551–62.

Ude, Donald Mark C. 2022. "Coloniality, Epistemic Imbalance, and Africa's Emigration Crisis." *Theory, Culture and Society* 39 (6): 3–19.

Ward, Patricia. 2021. "Capitalising on 'Local Knowledge': The Labour Practices behind Successful Aid Projects—the Case of Jordan." *Current Sociology* 69 (5): 705–22.

Ziółkowski, Mariusz, Jose Bastante Abuhadba, Alan Hogg, Dominika Sieczkowska, Andrze Rakowski, Jacek Pawlyta, and Sturt W. Manning. 2021. "When Did the Incas Build Machu Pichhu and Its Satellite Sites? New Approaches Based on Radiocarbon Dating." *Radiocarbon* 63 (4): 1133–48.

The Urgency and Benefits of Decentering and Decentralizing Knowledge Production

Knowledge from the Margins and the Social Studies of Ignorance

Several studies from the 1980s onward have documented cases in which knowledge has been decentered and decentralized, to use terms highlighted by the editors of this volume. Activist groups and broader social movements have advocated for the inclusion of nonscientists in knowledge making, given the stakes of these nonscientists in the knowledge produced, and that the knowledge produced by these nonscientists reflects novel perspectives, new methods, and new measures (see Kleinman 2000). In the course of their decentered and decentralized knowledge production, these groups have advanced their respective causes and reduced ignorance.

In this essay, I make the case that the failure to consider decentered and decentralized knowledge production in areas of broad social concern has led, or can lead, to prominent spheres of ignorance, and I argue that creating space for decentered and decentralized knowledge production allows for the possibility of more robust knowledge than would be arrived at without offering such opportunities. As markers for this discussion, I first highlight important instances of decentered and decentralized knowledge production. Thereafter, I explore the cases of AIDS treatment activism and clinical trial knowledge production and beekeeper on-the-ground knowledge production and the case of colony collapse disorder. In the cases I describe, the failure of mainstream institutions to take seriously decentered and decentralized knowledge led either temporarily or for the longer term to spheres of ignorance. Finally, I make a plea for institutionalizing opportunities for decentered and decentralized knowledge production, suggesting this broadens our understanding and reduces our ignorance.

Framing Terms

Before I get into the empirical cases, let me define several relevant terms. First, in this volume, Leandro Rodriguez Medina and Sandra Harding frame their discussion in terms of *decentering*, which they understand as thinking from the margins, edges, and the different. Harding develops this idea fully in her many books. In her volume *Sciences from Below*, Harding notes that "What we can know about nature and social relations depends upon how we live in our natural/social worlds. And peoples at the peripheries of modernity—women and other marginalized groups in the West and peoples from other cultures—have lived differently, with distinctive kinds of interactions with the world around them, than those at the centers" (2008, 5, 6). There is a social-structural element to this observation— one's social location (place in social [structural] relations) affects where one looks and what one sees. Bringing this down a level of abstraction, and to the margins in the United States, people with AIDS in the 1980s (see Epstein 1996), for example, had different interests in the research being done on AIDS than was typical of mainstream scientists. They saw and experienced things that mainstream scientists typically did not. They consequently sought answers to different questions about things like clinical trials than did mainstream scientists. They looked at the problem of clinical trial participation differently than mainstream scientists, and they proposed different measures of trial success and drug efficacy. In relation to mainstream scientists, they were on the margins and had distinctive experiences and so approached the problem of knowledge differently than mainstream scientists.

Similarly, the beekeepers that Sainath Suryanarayanan and I (2017) studied were on the margins of knowledge production about bees, despite the fact that their lives with bees gave them perspectives that mainstream scientists typically lacked. These beekeepers had a different experience of bee death and different understandings of how to think about honeybee health in beehives than mainstream scientists. In these cases, decentering knowledge would mean taking seriously the experiences and input of people with AIDS and beekeepers in seeking to understand AIDS and drug treatment, on the one hand, and honeybee health and solutions to high honeybee death rates, on the other. In the pages that follow, I consider how decentering can mean different things in different cases of knowledge production, and I contemplate the implications in each case for the kind of knowledge that is produced and, in parallel, the kind of ignorance that is reduced or not.

Let me turn now to decentralization. As Rodriguez Medina and Harding put it in this volume's introduction, decentralization means incorporating more actors from the margins in the knowledge production process than is typically done. I, as well as others, have written about this extensively (e.g., Kleinman 2000; Epstein 1996; Brown and Mikkelsen 1990) for the United States. There are many varieties of *incorporation*, and these, in turn, have implications for the kinds of decentering, the kind of knowledge produced. In short, decentering and decentralization are not, and cannot be, completely independent. One issue is whether this incorporation is institutionalized in mainstream organizations or reflects the persistent pressures of social movement organizations. Another issue is the type of incorporation involved. Are the actors from the margins involved in giving advice about research priorities, on one end of the spectrum, or are they making determinations about research methodology, on the other (see Kleinman 2000)? I discuss these different kinds of cases in the pages to follow.

I turn now to the problem of ignorance. At a most basic level, ignorance refers to not knowing something. Culturally, the term often has a derogatory valence, referring to a person's obliviousness, willful lack of knowledge, or cultural incompetence. Outside of common parlance, the easiest way to understand ignorance is by thinking in terms of Donna Haraway's (1988) notion of the god trick. This is the idea that we cannot see everything from nowhere. The world is infinite, and our perspectives are always and inevitably incomplete and situated. If we look one way, we see one thing. Look another way, we see something else. To oversimplify a bit for the sake of clarity, if we use a magnifying glass, we see the details of one leaf on one tree. If we use binoculars, we may see an entire stand of trees at a distance, understanding their distribution across the land but knowing nothing about individual leaves. As Gross and McGoey put it, following Alcoff and others, "ignorance is an inevitable consequence of the 'general fact' of our situatedness as knowers" (2015, 4).

Having for decades considered knowledge production, in the last decade of the twentieth century, scholars in science and technology studies (STS) began to consider the inverse of knowledge—what we don't see if we look a particular way with a particular tool (see Hess 2022). Among the relevant concepts here are undone science, selective ignorance, knowledge gaps, epistemic forms, and normatively induced ignorance.[1] In the simplest sense, *undone science* is scientific research that might conceivably be undertaken but is not. In this context, Kourany notes that ignorance and knowledge result from "pursuing certain lines of research rather than others" (2015, 155). However, there is considerably more sociological complexity here than this statement suggests. When Elliot (2015), for example, refers to *selective ignorance,* he describes the ways a company might collect information on the benefits of a product and not on the potentially harmful side effects. Similarly, a company might make choices about what kinds of information to disseminate and what not to. While a company might have the capacity to operate in this way, interested consumers might not be in a position to find out what is missing. They are ignorant, and resources, and thus power asymmetry, contribute to this ignorance.

Undone science points very explicitly to power and inequality. David Hess elaborates: "undone science draws attention to a kind of non-knowledge that is systematically produced through the unequal distribution of power in society" (2015, 142). Reformers and social movement activists seek research to answer questions of importance to their political goals, but find such research is not being undertaken. Some research is undone because funding agencies do not have an interest in a specific research agenda. Some is undone because it is technically and methodologically complex and so is arguably "undoable" (Hess 2015, 143; see also Frickel et al. 2010).

Scott Frickel offers the most fully developed idea of *knowledge gaps* in the context of his work on chemical exposure in the wake of Hurricane Katrina. Knowledge gaps or absences of knowledge result from "organizational practices, historical contingencies, and institutional arrangements" (Frickel and Vincent 2011, 12; see also Frickel 2014). Thus, if a government agency undertakes soil samples in a particular city, but only samples in particular precincts, there will be gaps in the knowledge we have about the missed precincts. If the agency only tests for particular chemicals, they will miss the possible presence of other toxic chemicals. In both cases, an absence of knowledge might result, and this absence might conceivably have implications for the health and safety of the communities not sampled and who might have been exposed to chemicals not tested for. The decentering of knowledge in this case and attention to the marginalized communities might lead to different testing and sampling protocols and so different knowledge.

Sainath Suryanarayanan and I (Kleinman and Suryanarayanan 2013) have developed two concepts that can provide insight into ignorance and its relationship to decentered and decentralized knowledge. First, we suggest that different knowledge producers and knowledge users, from scientists to government regulators to farmers and beekeepers, rely on distinctive *epistemic forms*. An epistemic form is a "suite of concepts, methods, measures, and interpretations that shape the ways in which actors produce knowledge and ignorance in their professional/intellectual fields of practice" (2013, 492). The epistemic forms on which scientists draw and depend are typically institutionalized and taken for granted. Thus, in a given field, it is acceptable to measure an object of study in a particular way, but not in others. Particular methods and technologies are acceptable for making a measurement, and interpretations are guided by the norms in the field.

Suryanarayanan and I looked at honeybee toxicologists, and we show that some methods used by these scientists and widely institutionalized and accepted by US regulators and agrichemical manufactures cannot address certain problems. The standards of evidence (what counts as legitimate—in this case statistically significant—results) used by these researchers lead them to ignore (statistically) inconclusive results, and their commitment to isolating individual variables leads them to ignore complexity and interactions. The result is knowledge that is incomplete and, indeed, arguably misleading. Put together, this situation produces *normatively induced ignorance*, our second concept, where the norms are deeply institutionalized and systematically lead the knowledge claims of beekeepers to be ignored or viewed skeptically.

Ignorance and the Value of Decentered and
Decentralized Knowledge Production: Two Cases

In this section, I highlight two cases in which the failure to consider decentered knowledge and to incorporate it into knowledge production discussions (to decentralize knowledge) leads to ignorance. Using these same cases, I show how creating space for decentered and decentralized knowledge allows or at least makes possible more robust knowledge.

I begin with the case of AIDS treatment activism made well known by the work of Steven Epstein (see, especially, Epstein 1996). Identified in the early 1980s, acquired immunodeficiency syndrome or AIDS was spreading rapidly throughout the globe, reaching in the neighborhood of 300,000 US cases by the early 1990s. The infection that causes AIDS, HIV, is spread primarily through unprotected sex, blood transfusions, and reuse of hypodermic needles. The rate of death among people with AIDS was quite high in the first decade after documentation of the disease, declining thereafter.

As outsiders to (on the margins of) the biomedical community in the 1980s, AIDS treatment activists, primarily gay men, were frustrated by the slow progress on treatment research for the deadly disease. Wedded to what was viewed as the gold standard for clinical trials, mainstream biomedical scientists adhered to the double-blind placebo approach to clinical trials. This meant that neither the scientists nor the patients knew who was getting the drug treatment under investigation and what proportion of the research subjects would get no treatment at all. At a time when a high proportion of people with AIDS were dying in the United States, this approach to research was ethically and practically unacceptable to AIDS treatment activists. Not knowing whether they were getting the placebo or the treatment drug, participants in clinical trials sometimes divided drugs/placebos among themselves, ensuring that all participants would get at least half the dose of the treatment under trial. Activists argued that the gold-standard approach to clinical trials was unethical and that all participants in clinical trials should get at least the best available clinical care. With personal stakes and a perspective more immediate and different than those of researchers, they proposed an alternative approach to clinical research.

While the gold-standard model has the potential to provide clean and unambiguous results, it can depend on the placebo-receiving group getting sicker or dying. It produces a particular type of knowledge about the drug under consideration, but it does not reveal how the drug compares to existing state-of-the-art clinical treatment. The result is undone research and normatively induced ignorance.

As Epstein (1996, 2000) notes, by the 1990s, AIDS researchers were taking the critiques of treatment activists seriously and considering the alternative clinical trial models proposed by activists. In an article in the *New England Journal of Medicine*, one researcher argued that all "limbs" of a given clinical trial should offer equal potential advantage to participating patients, that participants should not be denied treatment for opportunistic infections, and that trials should be altered or discontinued when it is clear that patients would do better with a different approach to treatment management. Decentering knowledge, taking seriously the understandings of the people who were directly affected by AIDS, led to significant methodological changes in clinical research. Undone science came to be done, since varied treatments were compared to one another, instead of one treatment being compared to a placebo. Furthermore, given the change in protocol supported by activists, researchers were better able to recruit clinical trial participants, leading to more robust research findings.

By the early 1990s, activists had been formally included in decision-making about approaches to research and treatment. Some became voting members on a crucial US National Institutes of Health committee (the AIDS Clinical Trials Group) and sat on hospital institutional review boards. As Epstein notes, by this point, activists "worked with scientists to determine the most profitable research directions, debate research methodologies, and allocate research funds" (2000, 15). The insights of activists on the margins were brought into the center, fundamentally changing the balance and types of knowledge and ignorance produced.

Getting to the point of decentralization—incorporating more actors from the margins—was by no means simple or automatic in the case of AIDS treatment research. The walls created by credentialed expertise are substantial. Most activists lacked formal science education, and they confronted established institutionalized research methods that those in the AIDS research community took to be the most effective and appropriate way to produce knowledge about AIDS and its treatment. The virtues of the approaches they used were taken for granted.

Again, Epstein compellingly describes the development of the AIDS treatment movement and the factors explaining their successful incorporation into decision-making in the AIDS research community. Treatment activists built on their experiences in the lesbian and gay movement of the 1970s and 1980s, and, as a result of that movement, gay men had existing organizations and structures they could utilize. Importantly, many of the activists were white, middle-class men in professional roles with political clout and the capacity to raise money. Buyers' clubs supplied people with AIDS with unapproved alternative treatments, activists published *AIDS Treatment News*, and some lobbied for alternative treatments. Importantly, many of these activists gained knowledge about mainstream research and learned to speak the language of researchers, giving them a kind of credibility that is rare for knowledge producers on the margins. This situation surely complicates our thinking about what we mean by being on the margins. The AIDS treatment activists were able to move from the margins toward the center in knowledge production. However, their movement was facilitated precisely because they were not on the margins in terms of social class and race.

The case of beekeeper knowledge making and the challenges it poses to mainstream research is more recent than the efforts of AIDS treatment activists. It shares some characteristics with the AIDS treatment case and is different in important ways as well. Like AIDS treatment activists, a group of beekeepers examined the work of mainstream scientists and found it wanting. Their on-the-ground experience—experience from the margins—suggested different ways than bee toxicologists to understand massive honeybee die-offs, and their efforts suggested novel methodological approaches to producing knowledge on what is killing honeybees. On the other hand, while there has been some acknowledgment of the thinking of this group of beekeepers, they lack an organized social movement, and the knowledge they produce has been largely ignored by the scientific community, government regulators, and the company that produces an insecticide considered to be an important part of the problem. The result has been normatively induced ignorance.

By the fall of 2006, beekeepers of all sorts were seeing their ostensibly healthy hives collapsing. Bees were abandoning hives and not returning. Beekeeping operations were losing between 30 percent and 90 percent of their hives, when losses caused by parasites, diseases, and poor nutrition had historically hovered around 15 percent. So-called colony collapse disorder (CCD) threatened and continues to threaten the sustainability of US agriculture. And this isn't because of what honeybee deaths could mean for honey production, although that isn't economically unimportant. More centrally, however, are the pollination services honeybees and beekeepers provide. In 2000, pollination through honeybees contributed nearly $15 billion to the agricultural economy in the United States. Farmers rent hives from beekeepers—some 2 million hives for more than fifty different crops, including almonds, apples, carrots, cotton, and blueberries (Suryanarayanan and Kleinman 2017).

The work toxicologists have done to understand CCD and the role of agrochemicals in contributing to honeybee illness and death reflects an extensive disciplinary and methodological history. It has been constrained by the epistemic form in which they work and largely taken for granted. Historically, in highly controlled laboratory and field experiments, scientists measure individual chemicals' lethal effects statistically. The idea is to understand individual factors and causal roles of these factors and to capture rapidly appearing lethal effects on targeted individual insects. Importantly, in stressing the rapid appearance of individual and lethal causal factors, the approach cannot consider the possible longer-term effects that could weaken honeybees without killing them. Equally, the approach fails to consider plausible interactions with factors beyond the specific insecticide of interest with other pesticides and pathogens. This approach, the epistemic form of bee toxicologists, cannot consider a complex set of interacting factors that might lead to slow, progressive effects over multiple generations in a beehive life cycle. Importantly, this epistemic form is also premised on a preference for conclusive evidence as defined by the statistical measure of 95 percent confidence that the results of an experiment are not due to chance. This is equivalent to preferring false negatives over false positives.

As Suryanarayanan and I note (Kleinman and Suyanarayanan 2013), this approach leads to three interrelated and overlapping types of ignorance. First, certain questions are not asked because established methods cannot address them. Thus, the work remains undone. Second, because certain questions are not asked, the results of the research are arguably misleading and distorted. Finally, by not taking seriously inconclusive results, scientists prefer a certain amount and type of ignorance over knowledge that is not unambiguous and virtually certain.

By contrast, some commercial beekeepers take another approach. They engage in everyday research in real time and in situ. It reflects their livelihood stakes. Their knowledge is built upon actual field conditions (not controlled laboratory or field conditions) and is attentive to the real-world complexity, not artificial simplicity, that bees and beekeepers experience. The informal analysis of beekeepers brings together multiple, complex aspects of honeybee colony health. Thus, these beekeepers examine brood pattern, the overall pattern in which a brood develops on a hive's comb. Brood pattern also provides beekeepers information on the local availability of nutritional sources for the bees and about queens' reproductive health.

The work of beekeepers qua researchers captures complexity. They conclude that there are likely a set of complicated interactions that adversely affect the health of honeybees. Unlike the application of an insecticide to a honeybee in a lab setting where scientists can see the dose that will kill a honeybee, in the field, beekeepers examine hives and see brood pattern deteriorating over time. With other comprehensive but relatively informal measures, beekeepers witness sublethal effects of a set of probable culprits (inadequate nutrition, parasites, insecticides, etc.). They speculate that these factors interact, possibly magnifying the effects of one another, harming honeybee health but not killing the bees immediately.

While there is increasing recognition that there is something to the claims made by beekeepers, bee scientists continue to utilize ultimately reductive methods, most likely to produce statistical certainty. Their epistemic form is widely accepted and its value and virtues commonly taken for granted. Their methods are recognized by colleagues and produce results likely to be published. Their approaches are accepted by the Environmental Protection Agency and Bayer, the company that produces the key insecticide at issue in the controversy over CCD. Bayer uses the status of the methods employed by bee scientists to argue that the chemical they produce is not the primary factor in hurting honeybees and should remain on the market. Along related lines, the EPA refuses to ban the substance. Of course, if beekeepers are correct, solving the problem of CCD would require reorganizing US agriculture, and that is even more unlikely than banning a single chemical.

The current state of affairs in the case of knowledge about honeybee health is that the only certified and accepted knowledge comes from established scientists. This is the knowledge used by federal regulators. While beekeepers have undertaken research and have suggested alternative interpretations of the conditions of honeybees, since the mainstream does not undertake research of the type informally undertaken by beekeepers, the research remains undone, there are knowledge gaps, and we face normatively induced ignorance.

Unlike the situation of AIDS treatment activists, beekeepers lack an established social movement. They do not have the organizational structures that would make it difficult for scientists, companies, and government regulators to ignore them. They lack political leverage. What is more, their complexity-oriented approach, which favors false positives or false negatives in terms of driving policy, amounts to a fundamental challenge to the well-established and widely accepted approaches to control-oriented, reductive science. We might say that knowledge here has been decentered, but since beekeeper knowledge has not been widely accepted, we cannot say knowledge in this realm has been decentralized.

Beekeepers have not had the success or impact that AIDS treatment activists have had. They have not managed to institutionalize a decentered approach. However, seeing promise in decentralizing honeybee research, in 2014 Suryanarayanan and I gathered a diverse group of stakeholders, including beekeepers of various types, farmers, and university scientists with different specialties, to facilitate collaboration and foster alternative research methods to investigate the complexity associated with honeybee decline (Suryanarayanan et al. 2018). Between 2014 and 2016, we undertook four full-day structured deliberations and interlaced these meetings with two field experiments centering on honeybee health. The group engaged in sustained interaction over time. The power of expertise was palpable in early meetings when beekeepers and farmers deferred to scientists. Over time, however, trust and respect developed in the group, and beekeepers began to speak up. In one case, beekeepers drew on their on-the-ground knowledge of cranberry pollination—knowledge from the margins—to explain sources of variability that scientists took to be noise. Beekeepers and growers stressed agronomic characters of particular cranberry marshes and proximity to honeybee colonies to explain variability in cranberry yield. Beekeepers' attention to smell, sound, and visual measures led them to propose important shifts in experimental design, as the group moved from the first to the second season of experimentation. Not surprisingly, the results of this collaboration across the boundaries between the margins and center of honeybee knowledge making did not produce decisive results and did not prompt a radical reconfiguration of honeybee knowledge making. At the same time, the collaboration arguably provides a template for future collaborations between scientists and nonscientists that could lead to a greater integration of research on the margins and greater decentralization in the center. It could offer the possibility of reducing ignorance and producing more robust knowledge and ultimately of the development of fairer approaches to complex phenomena of stakeholder concern.

From AIDS and Honeybees to Environmental Justice and Knowledge

In this essay, I bring ideas from the social studies of ignorance into conversation with the problems of decentering and decentralized knowledge production. Building on work from scholars like Sandra Harding and Donna Haraway, more recent research on the social studies of ignorance makes clear the importance of seeing ignorance as the flip side of knowledge and of understanding not just how and why knowledge is produced, but also why ignorance results. We must be aware of the implications of failing to produce knowledge—of producing ignorance—in areas of importance to marginalized actors. This is a key justification for decentering knowledge and decentralizing knowledge production. We can and must ensure that we produce knowledge not just from the vantage point of dominant social actors and with the tools of certified scientists, but that we take seriously the vantage point of people on different social margins and the knowledge that can be produced from their structural locations.

The failure to consider decentered and decentralized knowledge production can lead to ignorance of various sorts, and providing space for decentered and decentralized knowledge production allows for the possibility of more robust knowledge than would otherwise be possible. In the case of AIDS treatment research, while gold-standard clinical trials provided the possibility of clean unambiguous results on the efficacy of single drugs, the model led researchers to have trouble recruiting subjects and led participants to trade drugs in an effort to ensure nobody received no treatment. In this situation, the results could be less than clean, and under pressure from activists, scientists considered new clinical trial models, approaches that could compare subjects to themselves before and during treatments and approaches that compared different treatments, rather than treatment versus placebo. The effect was allowing research to get done, eliminating the ignorance that can result from undone science, and broader, more robust knowledge.

By contrast to the AIDS treatment case, the knowledge production of beekeepers went un- or underacknowledged by mainstream scientists, regulators, and corporate stakeholders. Knowledge produced by beekeepers was effectively ignored. Not considering attention to the complex interaction of factors over time and the sublethal effects of factors, as suggested by beekeepers, means an entire area of decentered research concerns and questions remains effectively unknown, and the dominant methods and measurement approaches lead to significant knowledge gaps.

There are many examples beyond the AIDS treatment and honeybee health cases in which critical decentered knowledge was produced and offered the promise of more robust understandings. Whether this research from the margins resulted in truly eliminating ignorance depended on the extent to which dominant social actors took the results of stakeholder research seriously, on whether knowledge production was truly decentralized. Let me describe two interesting cases.

Beginning in the late 1990s, Black residents of Norco, Louisiana sought "to demonstrate that the air that they were breathing was hazardous to their health" (Ottinger 2010, 245). Living near a Shell Chemicals plant, they were concerned about their exposure to toxic chemicals. To monitor the air, they used an inexpensive and easy-to-use technology known as buckets. Residents of broader fence-line communities founded the Louisiana Bucket Brigade. They were concerned about chemicals not monitored by regulators, undertaking health surveys documenting elevated rates of health problems, and were troubled about chemical exposure spikes. With regard to the latter, while regulators typically focus on average concentrations over long periods and compare these to ambient air standards, activists were concerned about air quality when it seemed especially bad. Thus, they monitored so-called flaring, accidents, and unplanned releases. They urged consideration of the ways in which occasional concentrated levels of exposure might contribute in specific ways to residents' illnesses. Focusing on annual average chemical exposures fails, residents contended, to consider whether occasional high-level exposures contribute in specific ways to residents' illnesses and whether periodic spikes have lasting health effects (Ottinger 2010).

It is clear that the thinking from the margins by members of the bucket brigade offers the promise of approaches to data collection not considered by those without the same kinds of immediate investments and concerns (e.g., regulators and chemical companies). Activists promote different kinds of sampling techniques and advocate sampling for a larger range of chemicals. They have revised (but have not transformed) established epistemic forms. Their work promises to reduce ignorance, filling knowledge gaps about exposure spikes and possibly chemicals not considered by regulators. That the buckets the activists use collect chemical residue in ways comparable to the techniques and technologies used by regulators gives activists' knowledge production some legitimacy. At the same time, because it deviates from established sampling practices, "the validity of bucket data is frequently questioned by regulators" (Ottinger 2010, 258). In sum, knowledge here has arguably been decentered, but not decentralized.

Activists in agricultural communities face problems not dissimilar to those in chemical industry fence-line communities. Those who live in agricultural areas can be exposed to the airborne movement of agricultural pesticides known as pesticide drift. Activists in some agricultural areas have worked with scientists from nongovernmental organizations to sample air for pesticides. They use a technology called a drift catcher, which, like the buckets used in Louisiana, draws on widely accepted technology to capture drifting airborne pesticides. In this case, activists do not necessarily seek to use different monitoring methods, but to monitor more fully. They argue that pesticide drift exposure is "rarely reported . . . frequently handled poorly by emergency responders, and . . . rarely investigated by doctors and/or local regulatory officials" (Harrison 2011, 705, 706). As Jill Lindsey Harrison notes, from the margins, local residents "can provide knowledge about the typical timing, types, and spatial patterns of pesticide applications in a particular region; knowledge about appropriate monitoring locations, and insights into local political opportunities associated with various pesticides" (2011, 708). While potentially filling an important knowledge gap, according to Harrison, the drift catcher technology is sufficiently complicated and the data sufficiently difficult to interpret and communicate that activists often cannot challenge regulatory-induced ignorance without NGO expertise and support. Here again, activists have decentered knowledge, using more or less established epistemic forms, but they have confronted other challenges to decentralization.

Conclusion

In all, my two primary cases, and the others I consider less fully, highlight the importance of power and structure in determining whether knowledge from the margins simply reveals ignorance (through decentering) or narrows it (through decentralization). These cases also capture intended and unintended consequences of seeking to bring knowledge from the margins toward the center and incorporate it. With regard to AIDS treatment, while activists were truly on the margins of knowledge production, they were able to offer insights that biomedical university and government scientists would simply not have come to. Their membership in the community of people with AIDS offered them insight. However, according to Steven Epstein, once activists were brought into government advisory groups and the like and began to have regular discussions with mainstream participants, they lost some of their perspective from the margin and began to think more like mainstream scientists and government officials. Here, we see the potential value of decentered knowledge, but also the problems in seeking to decentralize it.

The case of beekeeper research and CCD arguably adds a layer of complexity to the AIDS treatment case. While AIDS treatment activists called for a research design that diverged from the clinical gold standard, they did not object to established measures of efficacy and statistical significance. By contrast, beekeepers effectively challenged the possibility of isolating individual variables and claimed that one could not look at the health or illness of an individual bee but needed to examine an entire hive. They sought to capture and understand complexity at the cost of precision. In so doing, they effectively challenged the apparatus and organization of much academic science and regulatory decision-making, an established epistemic form. While some individual scientists did take beekeepers' approaches seriously (see Suryanarayanan et al. 2018), beekeepers' efforts did not lead to fundamental rethinking of bee research or pesticide regulation. The established epistemic form remained, and knowledge was not decentralized.

Bucket brigade activist pressure has led to policy change, but their efforts have not amounted to a fundamental decentralization of knowledge production. As Ottinger notes, government standards contribute to the structure of power relations and establish "the authority of scientists and other technical experts" and at the same time "marginalize alternative knowledge production processes" (2010, 251). Drift catcher activists face a challenge prior to decentralization. To be incorporated into the knowledge production process, activists need to produce knowledge. While those in rural communities often sense exposure to pesticide drift, the complexity of the technology sometimes limits their capacity to interpret collected data and so participate in debate.

Filling knowledge gaps and getting undone science done—reducing ignorance—depends on opportunities to decenter knowledge. Those on the margins—those directly affected—often can offer insights that could allow for new and better policies and practices. At the same time, these actors exist within an established set of power relations. Their knowledge is not just on the margins but is typically marginalized. They must challenge institutionalized academic disciplines and university structures as well as government regulatory bodies that commonly lean toward industry interests, and, finally, they must confront industries that are often invested in rejecting the knowledge produced by actors on the margins. Efforts by activists with substantial social and cultural capital, economic resources, and strategic and organizational capacities are more likely to be successful than those who lack these assets. Thus, simply highlighting the virtues of knowledge from the margins will never suffice. The successful integration of decentered knowledge—its decentralization—will depend on social movement mobilization and propitious social conditions.

Note

1   Interested readers can further explore research in the social studies of ignorance in *Routledge Handbook of Ignorance Studies* (Gross and McGoey 2015).

References

Brown, Phil, and Edwin Mikkelson. 1990. *No Safe Place: Toxic Waste, Leukemia, and Community Action*. Berkeley: University of California Press.

Elliot, Kevin C. 2015. "Selective Ignorance in Environmental Research." In *Routledge Handbook of Ignorance Studies*, edited by Matthias Gross and Linsey McGoey, 165–73. London: Routledge.

Epstein, Steven. 1996. *Impure Science: AIDS, Activism, and the Politics of Knowledge*. Berkeley: University of California Press.

Epstein, Steven. 2000. "Democracy, Expertise, and AIDS Treatment Activism." In *Science, Technology, and Democracy*, edited by Daniel Lee Kleinman, 15–32. Albany, NY: SUNY Press.

Frickel, Scott. 2014. "Not Here and Everywhere: The Non-production of Scientific Knowledge." In *Routledge Handbook of Science, Technology, and Society*, edited by Daniel Lee Kleinman and Kelly Moore, 263–76. London: Routledge.

Frickel, Scott, Sahra Gibbon, Jeff Howard, Joana Kempner, Gwen Ottinger, and David Hess. 2010. "Undone Science: Social Movement Challenges to Dominant Scientific Practice." *Science, Technology, and Human Values* 35 (4): 444–73.

Frickel, Scott, and M. B. Vincent. 2011. "Katrina's Contamination: Regulatory Knowledge Gaps in the Making and Unmaking of Environmental Contention." In *Dynamics of Disaster: Lessons in Risk, Response, and Recovery*, edited by R. A. Dowty and Barbara Allen, 11–28. London: Earthscan.

Gross, Matthias, and Linsey McGoey, eds. 2015. *Routledge Handbook of Ignorance Studies*, edited by Matthias Gross and Linsey McGoey. London: Routledge.

Haraway, Donna. 1988. "Situated Knowledges: The Science Question in Feminism and the Privilege of Partial Perspective." *Feminist Studies* 14 (3): 575–99.

Harding, Sandra. 2008. *Sciences from Below: Feminisms, Postcolonialities, and Modernities*. Durham, NC: Duke University Press.

Harrison, Jill Lindsey. 2011. "Parsing 'Participation' in Action Research: Navigating the Challenges of Lay Involvement in Technically Complex Participatory Science Projects." *Society and Natural Resources* 24: 702–16.

Hess, David. 2015. "Undone Science and Social Movements: A Review and Typology." In *Routledge Handbook of Ignorance Studies*, edited by Matthias Gross and Linsey McGoey, 141–54. London: Routledge.

Hess, David. 2022. "Undone Science and Social Movements: A Review and Typology." In *Routledge Handbook of Ignorance Studies*, 2nd ed., edited by Matthias Gross and Linsey McGoey, 167–77. London: Routledge.

Kleinman, Daniel Lee. 2000. "Democratizations of Science and Technology." In *Science, Technology, and Democracy*, edited by Daniel Lee Kleinman, 139–66. Albany, NY: SUNY Press.

Kleinman, Daniel Lee, and Sainath Suryanarayanan. 2013. "Dying Bees and the Social Production of Ignorance." *Science, Technology, and Human Values* 38 (4): 492–517.

Kourany, Janet. 2015. "Science: For Better or Worse, a Source of Ignorance as Well as Knowledge." In *Routledge Handbook of Ignorance Studies*, edited by Matthias Gross and Linsey McGoey, 155–64. London: Routledge.

Ottinger, Gwen. 2010. "Buckets of Resistance: Standards and the Effectiveness of Citizen Science." *Science, Technology, and Human Values* 35 (2): 244–70.

Suyanarayanan, Sainath, and Daniel Lee Kleinman. 2017. *Vanishing Bees: Science, Politics and Honeybee Health*. New Brunswick, NJ: Rutgers University Press.

Suryanarayanan, Sainath, Daniel Lee Kleinman, Claudio Gratton, Amy Toth, Christelle Guedot, Russell Groves, John Piechowski, Brad Moore, Deborah Hagedorn, Dayton Kauth, Heather Swan, and Mary Celley. 2018. "Collaboration Matters: Honey Bee Health as a Transdisciplinary Model for Understanding Real-World Complexity." *Bioscience* 68 (12): 990–95.

Making Difference
at the Edge

Conventionally in academia, we think of knowledge being made by knowing subjects whose epistemic privilege has been individually generated and
maintained through a series of certifications from disciplines and institutions, while being demonstrated by the circulation of original interventions in scholarly debates, all published in venues with access restricted by
rigorous peer review. I have studied those subject formations, the fashion-
ing and performance of the required discursive strategies, and the global
assemblages of highly stratified knowledge infrastructures, as well as how
they change over time. How and why do certain disciplines, institutions,
debates, venues, and infrastructures in certain places accrue the authority
of bestowing the privilege of abstraction, classification, and universality,
and others are saddled with specificity?

How is knowledge made at the edge of that assemblage of academic
power? What are the success strategies for making ideas and careers as
intellectuals while occupying the margins? Clustering at epistemic faultlines where conceptual ruptures are expected, those at the edge form robust meshworks, webs of relations for circulating their ideas and making
them more robust and resilient. Many, including Gloria Anzaldúa, Sandra
Harding, Manuel DeLanda, and Gregory Bateson, have taught us that embracing our mixtures, borderlands, margins, and differences can provide
an excellent standpoint for investigating and challenging the stylized assumptions, narratives, and practices of epistemic privilege, entitlement,
and authority.

I argue that we should study the practices at the edge of epistemic authority to understand how that authority is built, maintained, and sometimes
subverted. I claim that work is found in stories, gossip, whisper cultures,
and jokes, not in the formal, stylized templates of how to make and circulate

new knowledge. Furthermore, this chapter itself is not written within any of those conventional templates, with one exception: I have complied with the request to insert names and dates in parentheses within the text, as is the custom in some academic fields, but firmly rejected in others. That is but one example of competing customs about how, when, and if we are expected to refer to the work of our colleagues. Finally, I argue that various strategies for building and maintaining epistemic privilege, taken together, show the fragility of that authority, lessons for dismantling it, and guides for building a more robust, resilient set of epistemic practices.

The Middle Voice

Many academics think that language and discourse can be made into a neutral tool for representing something done elsewhere: making knowledge. Speaking and writing in the conventional academic way is performed to both claim and demonstrate that our work was done objectively, rather than subjectively, where all biases and carelessness have been neutralized. Telling stories, gossiping, and joking are seen as the domain of subjectivity and bias. Every quarter, some students tell me that while my stories are interesting, they wonder when we will get to work.

By contrast, some academics think that speakers and authors are never missing from texts and discourse; their putative erasures through use of the passive voice and avoiding first-person pronouns are simply devices to both obscure and enhance certain forms of agency, argumentation, and persuasion. In other words, the formulaic performance of neutrality can be the mark of power, authority, and privilege. As standpoint theorists have argued for decades, acknowledging agency and location in time and space is the beginning of a stronger form of objectivity, making explicit who is making claims, then explaining why and how. The focus is on process, the event of assertion, justification, and adjudication, not on the announcement of a product or findings. As Bradford Vivian states, acknowledging the act of "persuasion connotes a process rather than an outcome, a gerund instead of a noun. . . . Persuasion so conceived provides an account of thought, speech, and action that emphasizes processor relations instead of subjects and objects, difference rather than identity, ambiguity instead of transparency, and enactment rather than representation" (2004, 88–93).

Typically in English we think we are forced to choose between an active or passive voice and to choose between first-, second-, and third-person pronouns with all the associated cultural and epistemic connotations of those dyadic and triadic options. In many languages there is another option called the middle voice. The speaker and author can be reflexive, both actor and acted upon; the action can be both transitive and intransitive. The middle voice foregrounds the process of acting, making, and doing with an emphasis on explaining and justifying. On occasions of extreme importance, in matters of major concern the posture of objectivity, distancing ourselves, our bodies, and our minds, cannot be justified, intellectually or morally. I think we live in such times, defined by existential inequities, pandemics, precarities, toxicities, and disasters (Rubenstein 2003; White 1992).

For more than twenty years I have been trying to write in that middle voice, prompted by feminist epistemologies calling on us to reject the fallacies of dichotomies (like fe/male, wo/men, sub/objective, inductive or deductive, active or passive voice, etc.), along with my concerns about cultural, economic, epistemic, and social inequities. At first, I took on the task as a demanding intellectual exercise to sharpen my awareness of my analytic, interpretive choices. As I began to make presentations of my work in the middle voice, I was quite startled to discover that my interlocutors were quite offended and then shunned me. My claim that I had said or written nothing about my personal responses to the events and comments I reported was ignored. Speaking and writing in anything but the third person using the passive voice was considered subjective, confessional, and self-indulgent, independent of the content. It had to be condemned. For most, there was no alternative.

Those responses taught me that the conventional academic discourse was compulsory and violating that orthodoxy invited expulsion . . . or at least exile to the margins. My demotion from the status of someone seen as doing very interesting work to a subject of scorn was swift. Discipline and punishment are powerful informants. I began to reflect more on the epistemic status of voices and work found in the margins, including the stories of those pushed to the edge. Of course, that was not a new subject for me. My first book, also on knowledge-making communities, explored their stories, gossip, whisper cultures, and jokes. The shift was that I began to use the middle voice to juxtapose the complex epistemic practices in stories, gossip, whisper cultures, and jokes with those found in the more formulaic reports of knowledge making found in scholarly journals and books (Traweek 1988, 2021a, 2021b; Sørensen and Traweek 2022).[1]

What Is Epistemic Privilege?

I would argue that epistemic privilege is the repetitive performance of a series of compulsory figures, much like ice skaters displayed in Olympic competitions until 1990. Those of us with epistemic privilege in academia are expected to speak and write with certain formal vocabularies and grammars. To claim our professional identity, we must erase our personal agency by using the passive voice and avoiding first-person pronouns. We must display our commitment to abstraction by the repetitive use of singular generics. We must write about our research using a formulaic template in use for hundreds of years: a statement of our research topic, questions, hypotheses, methods, interpretive strategies, and findings. We must also know what topics, questions, hypotheses, methods, interpretive strategies, and findings to ignore. In presenting our own work, we must specify precisely what we have retained and challenged in the earlier scholarship on our research topic and why. We must display our capacity for categorical thinking, identifying precisely what is included and excluded from our classification systems and why. We must be skilled in both inductive and deductive reasoning. We should specify our enabling assumptions. We must explain precisely what we mean and do not mean. We must maintain an authoritative posture, holding our bodies in certain ways and projecting our stylized voices firmly. We must do all that with precision and facility, all while occasionally incorporating the locally sanctioned forms of wit, which might be irony, sarcasm, or puns. It is embarrassing to report that it has taken me decades to learn that my own preference for dry ironic wit is not widely shared among my colleagues, and my own efforts usually have been taken literally, with unpleasant consequences for me.

Research in feminist epistemology has challenged some of those practices. Investigating the binary classifications of gender and sexuality has led many feminist scholars (but clearly not all) to examining the assumptions embedded in all binaries and any classification system, inquiring what parts of the spectrum of possibilities are to be excluded, erased, and marginalized, to challenge the law of the excluded middle. That includes the stark juxtaposition of subjectivity and objectivity, body and mind, self and society, nature and culture, metropole and periphery. Questioning the singular generics of *he* where men stand as synecdoche for all people has led many (but clearly not all) to exploring the assumptions embedded in all singular generic abstractions and universals, from beauty, truth, and logic to sex, race, and class. Challenging the assumptions embedded in patriarchies leads most feminist scholars (if not all) to challenge the presumptions in all exceptionalisms. Studying how the concepts of man and woman, male and female have been constituted as mutually exclusive provokes for most (if not all) questions about all forms of essentialism. Clearly, there are some who self-identify as feminists who are not students of feminist epistemology; they are comfortable with constituting definitions that sort people into identities, generics, classifications, essentialisms, and exceptionalisms. Of course, some do that strategically, imagining that temporarily coalescing a group as exceptional is a useful way to undermine the privilege of other groups that justify their own power with essentialist and exceptionalist identities. I think not. As Audre Lorde (2007) admonished, the master's tools cannot dismantle the master's house.

Mapping Imperial and Colonial Epistemic Privilege in Global Academia

As an undergraduate history student in the 1960s and an anthropology graduate student in the 1970s, I was taught to always be wary of the usually unexamined assumptions in research embedded, however indirectly, in colonial, imperial, wartime, and development agendas. In some parts of anthropology (but not all) there is a rather long history of concern about how imperial political economies have shaped even the most arcane scholarship about the world. Of course, questioning the imperial assumptions and condescending implications in much scholarship about the world became more common after the appearance of books like Said's *Orientalism* (1978) and Chakrabarty's *Provincializing Europe* (2000), at least in some fields. That is, I learned to study how scholarship is situated in global political economies and often bears the mark of its time and location in not only its subject matter but also the structure of its research questions, hypotheses, modes of investigation, and interpretive approaches (Traweek 2000).

The point is that much conventional research does not address the "orientalist" and condescending assumptions that might shape any work conducted within empires. For centuries, bright minds and good data have been extracted from the so-called periphery and brought to the imperial centers for what is termed value-added education and analysis, benefiting the imperial metropoles, just as so-called raw materials have been extracted from colonies and brought to the center for assembly with other resources, then recirculated as far more valuable commodities than their original components. Appropriation and commodification in regimes of circulation are part of knowledge making as well as other goods and services. What is surprising is that many historians, sociologists, anthropologists, and so on of science, technology, and medicine, as well as those in the fields of science, technology, and society (STS) studies and women's, gender, feminist, and sexuality (WGFS) studies still presume that important knowledge, including in their own field, is diffused in one direction: from the current and former imperial metropoles to the peripheral rest. I have often heard people in both fields say that important theoretical work is still done primarily by those based in the metropole (whatever the scholars' migration story might be) while useful information about "cases" is brought from all the peripheries to be interpreted in the metropole. Everyone has a role to play in that hierarchical structure, including at the edge.

I have found the critique of "orientalist" STS or WGFS to be part of a whisper culture rather than a more open discourse about the important scholarship at the edges of those fields that is provincializing the work done in the metropoles. That gap is the subject of this essay. One part is the hard work of maintaining conventional epistemic privilege and authority. As we all know, maintenance work is necessary to keep the simplest apparatus operational or the smallest dwelling tidy. Intriguingly, much of that important work is routinized and done by a specialized, but underrecognized, heavily surveilled labor force. In academia, that exacting maintenance work is done by those who have learned to meticulously follow established patterns. Compliance is rewarded, and deviation is punished. Nonetheless, new kinds of work have emerged.

Studying Many Ways of Knowing with Multiple, Overlapping Epistemic Practices

A set of practices conventionally called Western science has been juxtaposed with other ways of investigating the phenomenal world, presuming that the former is far more robust than the latter. Upon closer examination, it became clear there are many flaws in that argument. For at least sixty years, feminist epistemologies also have challenged the notion that certain kinds of knowing, such as those used in the science, engineering, and medical fields, are essentially and exceptionally more authoritative than others. Many others have challenged that positivist position for a century. Together, anthropologists, historians, sociologists, philosophers, and many other kinds of scholars have established that ways of knowing, privileged or not, are all historically situated in global and local political economies, including imperialism, colonialism, wartime, and peacetime, as well as being embedded in the different locally specific forms of capitalism, communism, and socialism. Over time, the adjudication of knowledge claims in those fields follows patterns called paradigms, epistemes, thought styles, climates of opinion, worldview, Weltanschauung, epistemic cultures, actor networks, and so on. We have studied how research fields emerge, coalesce, gain ascendancy, and wane. We have learned the ways practitioners amass resources and transmit knowledge across generations; we have studied how practitioners, questions, and modes of inquiry circulate, plus how they adjudicate knowledge claims among themselves (Felt et al. 2018; Harding 1993, 1998, 2011; Restivo 2005; Biagioli 1999).

Alongside studying those practices among scientists, engineers, and physicians, we also study many other kinds of knowledge-making communities. Extensive bodies of research have emerged on these multiple ways of knowing: the different research fields investigate what they call ethnosciences, Indigenous Knowledge Systems (IKS), local knowledges, polycultural knowledges, and traditional knowledges, as well as the fields known as science, technology, and medicine. Sandra Harding (1998, 2008), one of the editors of this volume, long ago argued that all knowledge is local, that all sciences are ethnosciences; she also has recommended that we attend carefully to "sciences from below" such as the knowledge made by "women in households" who carefully assemble overlooked resources to accomplish complex tasks strategically.

Note that the field of IKS has long focused on the rich and subtle knowledges generated by indigenous people around the world. In those ecologies, much important knowledge and knowledge-making practices have been transmitted orally across generations. There has been extensive research over the last century attesting to the robust, resilient features of oral traditions; part of that is accomplished through employing complex structural strategies embedded in narrative forms.

Investigations of the multiple epistemic practices, from university based to community based, show extensive overlap and find similar patterns. We also study how those groups coexist with each other, appropriate each other's knowledge work, and compete for resources. Often they are arrayed in hierarchies of epistemic authority; typically the scientists, engineers, and physicians are the most privileged, even when there is dense traffic among them. One consequence of that work on global situated learning in different communities of practice has been a set of initiatives from the United Nations, the World Bank, and the World Intellectual Property Organization to guarantee that indigenous groups get patent rights for their extensive knowledge based on materials and techniques for ameliorating illness and facilitating wellness, knowledge that had long been investigated and appropriated without acknowledgment by university-based researchers.[2]

Seeking New Epistemic Practices at the Edge

If we pursue feminist epistemologies, eschewing binaries and classifications, essentialisms and exceptionalisms, how do we find knowledge being made at the edge of those multiple knowledge-making communities? How do practitioners, questions, and modes of inquiry circulate at the edge? How do new lines of inquiry emerge and change? How do those epistemic cultures work when they do not use the master's tools?

Over the last several years, some members of the Society for Social Studies of Science have launched impressive projects on "transnationalizing STS," a very welcome and important widening of both perspective and awareness of the many kinds of STS research being conducted around the world. At about the same time, transnational WGFS studies have gained much greater attention. Conventionally, most researchers in the STS and WGFS fields have focused on a limited set of locations, clustered in global power centers. The remaining places studied have tended to be part of the empires of that limited number of power centers, but peripheral to the controlling metropoles. For example, there is much research by European-based scholars about places that have been part of the European empires, just as there is much research on places that have been part of the Chinese, Japanese, Persian, Roman, Russian, Soviet, US, and other empires by scholars educated and based in those powerful countries.

In contemporary wealthy, privileged knowledge-making communities, exemplary knowledge usually is stored in data banks, archives, and museums. Most have extensive metadata records about the knowledge artifacts, including their provenance, the distinctive genealogy of authorship and ownership of a particular artifact. Those exemplary artifacts are compiled to discipline our gaze, focusing our attention. The textbooks, encyclopedias, and so-called high-impact journals establish the controlling narrative of powerful work, maintaining the authority of dominant lines of inquiry. Standardized forms of knowledge production are reinforced, while innovation emerges within strictly proscribed limits. Disciplinary pedagogies reinforce what qualifies as important questions, modes of inquiry, and interpretive approaches, as well as what is to be ignored. One physicist mentioned that when he raised certain sorts of questions, his teachers told him to take philosophy courses, not physics. Another said, "if I did not want my ideas tested by nature, I would be a mathematician." Anthropologists, historians, and sociologists make similar remarks about what is not in their fields. There are extensive injunctions in each field about what must be ignored in becoming a quali-fied practitioner. Many students circulate jokes about those prohibitions; a database of more than 2,500 such jokes is maintained by Joachim Verhagen.[3] The humor reinforces the enabling assumptions of the disciplines.

However, even in university-based communities of knowledge mak-ing with strict disciplinary pedagogies, much crucial knowledge circu-lates without a clear notion of its origin or histories of its pedigree. Much knowledge about the daily processes of making and changing knowledge is sometimes called corridor talk, because it is rarely written and yet widely discussed outside the formal venues of seminars and lectures. Why is there a dyadic juxtaposition of a highly articulated, disciplined canon and a rich, informal oral tradition? What is preserved in each and why?

For example, many claim that the "gold standard" in research is replica-tion, meaning that research is reproduced in order to see if the same results are achieved. However, little research is replicated, in part because it rarely would be funded and in part because refutation is rarely a career-building strategy. Furthermore, only those with the embodied memory of that par-ticular research experience would know how to replicate the process. In practice, knowledge makers corroborate: looking to see if the same answer emerges when using different resources, methods, theories, and labs.

I argue that the formal written record is open to revision, albeit according to strict rules of engagement. What is not written but discussed endlessly is a firm set of community beliefs about significance, quality, style, stan-dards, ethics, aesthetics, and reputation. Those beliefs are not under re-vision. Novices learn through the jokes and stories what to admire and scorn, plus how to tell the difference. Adepts use the stories and jokes to adjudicate reputations for quality and trust. I was told long ago that just because something was published in an important journal does not mean physicists should pay attention to it. No outsider, no matter how well read in the published scholarly literature of each field, would know how to pro-ceed without acquiring the accumulated, daily embodied experience from the fully initiated knowledge makers themselves. They know the procedures, equipment, and data that are always under revision. Those are the storytellers and the jokers; it takes years to become a skilled practitioner.

Between the announcement of new, interesting work and some taking on the work of corroboration, there is a vast discussion about the quality of the first work, given that all the obvious standards were met. I have learned that researchers constantly reflect on whether or not they, in their expression, "trust" the work of their colleagues. Several years ago, in presenting my work to some physicists, I mentioned that an American sociologist had argued that trust was necessary only in emergent research communities; once fully established with strong organizational resources, an important, established research field could rely on resilient administration and formal peer review, instead of building webs of trust. The physicists laughed. When I told them that research had been published by MIT Press, they were shocked. They said everyone should know that argument was profoundly wrong, even dishonest. Practitioners know that the processes for making even the most privileged knowledge is not codified, and they know that research communities require resilient webs of trust.

Where is the archive of embodied memory for making new kinds of knowledge? Where are the guidelines for building trust in the work being done? Where are the sanctions for violations of community ethics? They exist in the vast trove of stories, gossip, whisper cultures, and jokes that circulate among all resilient communities, many of them generated and retold by the people at the edge of privilege. That informal knowledge exists alongside the institutional assemblage of resources in which academic research communities are embedded: universities, laboratories, departments, funding agencies, research journals, scholarly societies, and so on, whatever their rank in the epistemic hierarchies of privilege.

That is, in the world of university and individual metric rankings, the quantity of research grants from certain agencies and research publications in designated high-impact journals certainly counts. However, prominent senior scholars in many fields lament the decline in quality of research in their fields as metrics culture increasingly prevails. Recently at a conference panel, I noted that UCLA was increasingly ranked equal to or higher than UC Berkeley; a copanelist laughed and expressed incredulity. I explained that Berkeley's accumulated, tacit reputation did not diminish its long-term decline in the metrics of research funding, student applications, and so on. That is also the case with several private Ivy League universities in the United States, as well as specific departments at those schools.

Nonetheless, in the study and practice of research, the conventional epistemological status of stories, gossip, whisper cultures, and jokes is nil. I beg to differ. It is in that discourse that knowledge is made and adjudications about quality are reached. Only then is recognition achieved, artifacts and ideas inserted in museums and textbooks, and a few canonized. Access to all that talk is highly restricted. Traditionally, building reputations required access to the corridors of the labs and university departments where that discourse has flourished. It would have been nearly impossible for someone without that access to make major contributions to a conventional research field. In the past, access to the recognized funding and publishing venues required prior access to the informal discursive domains. How have those borders performed?

Building Trust outside the Classifications

In *Improvised News: A Sociological Study of Rumor*, Tamotsu Shibutani (1966) argued that gossip was a strategy for exploring the meaning and truth of ambiguous information. It was also a strategy for negotiating reputations when the community chooses not to rely on formal institutions to arbitrate transgressions. Knowledge-making communities use gossip to adjudicate the meaning and resilience of knowledge claims, as well as circulating news about those who transgress the community ethics. Since the 1970s, I have listened to physicists and astronomers discuss whether or not they trust the data in new knowledge claims and whether they trust their colleagues to behave honorably. They also talk about how to make new claims themselves and how to boost their own reputations. They discuss the quality of everything. I study the content, structures, and circuits of their stories. In this context it is important to note that the template for research funding proposals and journal articles is not the narrative template for trust in colleagues and knowledge claims. The formulaic public template is a required place marker that must be generated and performed well. Similarly, the narratives about trust and ethics must be generated and performed well. The actors in the two scripts can be from very different subject positionings in the community.

Why are there these two different genres, the template for research funding proposals and journal articles, and the narratives about trust, ethics, reputations, and quality? I argue that the standard template can be debated in public according to certain conventional discursive criteria.

Access to open participation in those debates is restricted, and that restricted access usually is controlled by prominent, powerful members of the research community who also control access to important resources. The speakers and audience for stories about trust, ethics, reputations, and quality are everywhere; many are in marginal positions. The powerful ones worry about the stories. Many years ago, I heard about an internationally known scientist who had a list of rules prominently posted in his lab; "no joking" was on the list. He was known for working very hard to control the circulation of informal news about his lab to the rest of the community. Was the story itself a joke? That story circulated on several continents, along with many more tales about the lab, its data, and its work processes, all focused on extreme control. The public persona was very strong; the global gossip circuits told a very different story. Different kinds of power were enacted through the two different communication practices. Everyone understood the two kinds of discourse; we find that each explicates the other. Tracing them both, we learn those at the edge can revise reputations of even the most powerful, but access to major resources is still limited to those who control the apparatus of formal communication practices and templates. Nonetheless, the opportunity to share knowledge at little or no cost is gradually dismantling the gates of privileged communication practices.

The relationship between the two genres and their conventional domains is being revised. Stories of abuse and bullying circulate globally in academia. During the past decade, some stories about harassment in academia have been moved from whisper culture stories into public discourse by leaders in the field. Reports of harassment in astronomy have led to powerful people losing their access to and control of significant resources. One university department organized a letter to the *New York Times* from 228 scientists in ten countries protesting the newspaper's reports on one incident because it sided with the perpetrator rather than the victims. One academic society announced on its website that tacitly condoning harassment constitutes a failure in senior leadership and should be reported in evaluations of them (Overbye 2015; Diniega et al. 2016).

At the Edge and beyond the Walls

Constructing conventional epistemic authority requires building borders and walls, then defining who and what goes on which side of those categories. A few qualified applicants will be invited in and tutored in how to become the kind of subject who wants to think and write as they should. How that edifice has been built, maintained, and expanded globally over the last several centuries was beyond the scope of this chapter. The point here was to chronicle its strategies for enforcing its epistemic authority and privilege. Its powerful forms of surveillance and disciplining have created a very large, steadily growing global demographic of the marginalized, exiles, and aliens.

The ever-expanding, complex exclusionary practices of surveillance, evaluation, disciplining, and punishment have been instructive experiences for all of us. Some at the edge and beyond sadly replicate the surveillance, classificatory, and exclusionary practices of the most privileged. We now have a paradox: the accumulated reputations of certain universities, departments, fields of inquiry, conceptual strategies, and discursive practices once determined access to privileged funding and publishing venues; in some places, like the United States, those accumulated reputations no longer have that privileged access. In fact, the upstarts like Texas A&M, UCLA, UC San Diego, and University of Washington have had greater access to those privileged venues for about two decades; their current, sustained metric rankings exceed those of the traditionally privileged. We can ask how that has happened, which Knut Sørensen and I have done in *Questing Excellence in Academia* (Sørensen and Traweek 2022); we also can ask how that gap has affected those of us in the margins of either kind of privilege.

The rest of us can and do benefit from being at the crossroads of so many interesting journeys; we can share our road maps of what not to do. Many of us in the borderlands and those beyond the walls and categories of privilege are exploring new ways to build knowledge carefully without constructing yet more epistemic privilege, surveillance, and exile. There are many of us thinking analogically with more abduction and less in/deduction. We are making many kinds of stories, remixing our global tropes in new ways. We are gathering, communicating, and circulating our work in newly created unranked venues, some of which do not require costly travel or subscriptions, invoking new technologies.

Many of the risks of living in the margins have been experienced, protested, and documented for centuries. Are there new risks for those choosing to work in the gaps created by declining and rising privilege, between the Berkeleys and the UCLAs, between their rising and declining access to the same resources, as we seek new resources, circulating our work and our own rumors with new communications technologies? Those in power are still too distracted by their conflicts with each other to give us much attention or be bothered by us. However, while they are at war, we are creating new kinds of play. As we accumulate newer kinds of reputations, we are a growing source of annoyance. As wealthy fields of inquiry, their funders, and their publishers debate how long to embargo new research data, a new research device, formerly named the Large Synoptic Survey Telescope and now known as the Vera Rubin Observatory, aims to collect and immediately deposit data on the internet. It is, as it were, a new business model for making knowledge, much like the Grateful Dead becoming well known even as they encouraged fans to make and circulate their own recordings of the band's concerts. How will these new political economies of new ways of knowing fare if the warring great powers fade, as they usually do? Are we migrating marginals at risk of becoming the next new regime? I think that partly depends on our responses to the conventional enticements to render our epistemic margins into centralized orders, hierarchies, and universals. At such moments, I ask myself: What would the labor leader and civil rights activist Dolores Huerta do?[4]

We are performing our curiosity with journeys through places and practices with palimpsest maps, layered like mille-feuille. The folds' crests and crevasses lead us to new borders and faultlines. We search for patterns in the excluded, in the anomalies and noise. Through our epistemic travels, we form meshworks of collaborators where new work is shared more than it is owned. We have polyglot ways of circulating and discussing our work. In the midst of the powerful resources of the borderlands and beyond the walls of epistemic privilege, I am suggesting that we fully abandon singular generics, binaries, classification systems, essentialisms, and exceptionalisms. Those once-privileged epistemic practices undermine our new kinds of research design, modes of inquiry, interpretive approaches, pedagogies, and engaged social action. We can practice new epistemologies at the edge.

Back to the Middle, betwixt the Edges of the Categorical Imperatives and Normativities

A reviewer asked me to elaborate on the narrative strategies I am using in this chapter; more specifically, a reader asked that I explain, for those who might be shocked, that at least in some parts of the social sciences, writing about our research in the first person with active verb forms has been done by respected scholars for at least a half century, for excellent epistemic reasons. I was reminded that in every class I teach, I hear otherwise very smart students reporting that they want to use evidence in their reports, not an-ecdotes, and when pressed, they clarify that they want to use numbers instead of words to support their arguments, explaining they want their work to be without bias. We all have heard those mantras many thousands of times, and most learned long ago to recite them as a catechism. Here I want to call attention again to the binary and hierarchical classification of evidence and anecdotes, numbers and words, passive and active voice, third- and first-person pronouns.

Of course, I should not be surprised by how many intelligent people will resort to such catechisms that have no foundation in reason. Clearly, most scholars have learned to practice what has been preached, rather than stopping to think carefully about what their recitations mean. I ask those students some Socratic questions, as I have for decades: what do they mean by evidence and anecdote, why do they see those terms as dichoto-mies, and why do they think numbers and words can be sorted into those two categories. I ask them if they can think of numbers and formulas that are anecdotes, words and narratives that are evidence, and vice versa. I ask if they know how to mislead subtly with formulas, use narratives truthfully, and vice versa. I try to teach them how to build complex, robust arguments without resorting to simplistic, formulaic narrative templates that signal so-called objective discourse. I ask who and what benefits from defining objectivity simply by the use of numbers, specific pronouns, and certain grammatical forms. If objectivity is so important, why are the criteria set so low and kept there by so many for so long?

Over the last half century, feminist scholars and activists have chal-lenged binary, hierarchically ordered gendered social classification systems, increasingly arguing for a more fluid spectrum of possibilities, as found empirically around the world in many life-forms. Similarly, biologists have found that the various racial social classification systems do not correlate with the more fluid spectrum of genetic variations or the different, diverse clusters that appear when we focus on certain human characteristics. Simi-larly, feminist epistemologists have challenged the assumptions embedded in many other binary classifications, such as those arbitrarily differentiat-ing bodies and minds, subjectivities and objectivities, humans and natures, qualitative and quantitative research methods, and the narrative templates that require we never use first-person pronouns or active verb forms, as if using those terms and grammars showed we were incapable of thinking clearly or doing robust research, as I have learned that many still believe and, worse yet, want to enforce. I have learned to live with being labeled by some as undisciplined; I prefer the epistemic edges where the routine performance of orthodoxy is questioned, rather than vice versa.

Feminist epistemologies challenge the assumptions embedded in social categories of races, genders, classes, sexualities, ethnicities, dis/abilities, precarities, inequities, ages, nationalities, and so on, examining more carefully what imposing such categorical thinking does to our capacity to investigate, learn, and understand. While classificatory thinking clearly can be a temporarily useful heuristic device (like a thought experiment), in this chapter I have argued that what it conflates, essentializes, and obscures should never be forgotten, intellectually, personally, ethically, politically, and epistemologically. Certainly I am well aware that some scholars want simultaneously to challenge gender binary classifications and to enforce epistemic dichotomies in scholarship. I have found that disabling paradox to be a troubling signal of other fundamental flaws with large consequences; I am exploring that issue elsewhere. Here it should suffice to note that it is far more demanding to live, work, and think in the borderlands where we must continuously interrogate and debate what we mean and why, examining our assumptions outside the orthodox classifications and categories, in order to build the robust knowledge we need to dismantle the entangled epistemic, material, social, and ethical inequities. Simply sorting numbers and words, verb forms, and pronouns is not up to the task.

Finally, for centuries some have argued, as feminist, indigenous, queer, and crip epistemologists, among others, have for many decades, that there is a wide range of excellent, embodied, experiential knowledge practices and logics, some borne of managing to live effectively in the margins of domination and oppression. We are learning how to articulate those positively, not merely cataloging how the dominant knowledge practices and logics systematically, routinely, and violently exclude them, from massacres to the thousand cuts of daily microaggressions and insults, from punishment to disciplining. Those excluded practices and logics thrive in complex and subtle expressive forms, narrative genres, and logics. Elsewhere I have written of my own ongoing subject formation in academia and the transgressive corridor talk that has sustained me as I've slowly learned to understand the reasons for all the epistemic abuse I have witnessed and endured. As many of us find common cause and together envision making knowledge differently, some of that knowledge circulates as stories, jokes, songs, and dances, composed as we turn our difficult experience into artful expression. I began learning more about the middle voice and many others not found in the dominant narratives when, as a high school student, I would sneak into the Lighthouse Café for jazz, quickly realizing that some knew how to make new kinds of knowing by reimagining all the rules; later I found more experimentation at the Jabberwock. Some knowledge emerges in actively voiced first-person forms; this one is written in the middle voice, laced with the irony braided from pain and joy, and it has benefitted enormously from several important interlocutors, some of whom include Kim Fortun, Mike Fortun, Jade Vu Henry, Ariel Hernandez, Jarita Holbrook, Karin Knorr, Nashra Mahmood, Martin Oliver, Kyriaki Papageorgiou, Maria do Mar Pereira, Zoe Safoulis, Chela Sandoval, Knut Sørensen, Alluquere Stone, Hirotaka Sugawara, and Nadine Tanio, plus the editors of this volume.

Notes

1   See also my blog (Sharon Traweek, https://corridortalk.blog/corridor
-talk/) and these interviews of me: "Oral History Booth 2018 Sharon
Traweek" (Klett and Schütz 2018) and "Searching for How Epistemic
Power is Made" (Traweek, Kaşdoğan, and Fortun 2021).

2   In 2007 the UN placed intellectual property rights on its list of human
rights for indigenous people: "States shall provide redress through effective mechanisms, which may include restitution, developed in conjunction
with indigenous peoples, with respect to their cultural, intellectual, religious and spiritual property taken without their free, prior and informed
consent or in violation of their laws, traditions and customs" (United
Nations 2007). See also World Bank Organization 2010; World Intellectual Property Organization (WIPO) 2024. WIPO Intergovernmental
Committee Session documents are in the WIPO archives: https://www
.wipo.int/meetings/en/archive.jsp.

3   See Verhagen's Science Jokes website, https://jcdverha.home.xs4all.nl
/scijokes/index.html.

4   See Grateful Dead (https://www.dead.net/) and Dolores Huerta Foundation for Community Organizing (https://doloreshuerta.org/).

References

Biagioli, Mario, ed. 1999. *The Science Studies Reader*. London: Routledge.
Chakrabarty, Dipesh. 2000. *Provincializing Europe: Postcolonial Thought and
Historical Difference*. Princeton, NJ: Princeton University Press.
Diniega, Serina, J. Tan, M. S. Tiscareno, and E. Wehner. 2016. "Senior Sci-
entists Must Engage in the Fight against Harassment." Eos, September 8,

2016. https://eos.org/opinions/senior-scientists-must-engage-in-the
-fight-against-harassment.

Felt, Ulrike, Rayvon Fouché, Clark A. Miller, and Laurel Smith-Doerr, eds.
2018. *Handbook of Science and Technology Studies.* Cambridge, MA: MIT
Press.

Harding, Sandra, ed. 1993. *The "Racial" Economy of Science: Toward a Democratic Future (Race, Gender, and Science).* Indianapolis: Indiana University
Press.

Harding, Sandra, ed. 1998. *Is Science Multicultural? Postcolonialisms, Feminisms, and Epistemologies.* Indianapolis: Indiana University Press.

Harding, Sandra. 2008. *Sciences from below: Feminisms, Postcolonialities, and
Modernities.* Durham, NC: Duke University Press.

Harding, Sandra, ed. 2011. *The Postcolonial Science and Technology Studies
Reader.* Durham, NC: Duke University Press.

Klett, Joseph, and Tim Schütz. 2018. "Oral History Booth 2018 Sharon
Traweek." Video interview. STS Infrastructures, Platform for Experimental Collaborative Ethnography. https://stsinfrastructures.org/content
/oral-history-booth-2018-sharon-traweek.

Lorde, Audre. 2007. "The Master's Tools Will Never Dismantle the Master's
House." In *Sister Outsider: Essays and Speeches,* 110–14. Berkeley, CA:
Crossing Press.

Overbye, Dennis. 2015. "Geoffrey Marcy, Astronomer at Berkeley, Apologizes
for Behavior." *New York Times,* October 11, 2015.

Restivo, Sal, ed. 2005. *Science, Technology, and Society: An Encyclopedia.* New
York: Oxford University Press.

Rubenstein, Richard L. 2003. "Review: Writing History, Writing Trauma."
Holocaust and Genocide Studies 17 (1): 158–61.

Said, Edward. 1978. *Orientalism.* New York: Pantheon.

Shibutani, Tamotsu. 1966. *Improvised News: A Sociological Study of Rumor.*
New York: Irvington.

Sørensen, Knut H., and Sharon Traweek. 2022. *Questing Excellence in Academia: A Tale of Two Universities.* London: Routledge.

Traweek, Sharon. 1988. *Beamtimes and Lifetimes: The World of High Energy
Physicists.* Cambridge, MA: Harvard University Press.

Traweek, Sharon. 2000. "How Modern Became Retro: An Historical Political
Economy of Knowledge." In *Doing Science + Culture: How Cultural and
Interdisciplinary Studies Are Changing the Way We Look at Science
and Medicine,* edited by Sharon Traweek and Roddy Reid, 21–48.
London: Routledge.

Traweek, Sharon. 2021a. "I Prefer the Map." *Engaging Science, Technology, and
Society* 7 (2): 56–64.

Traweek, Sharon. 2021b. "Let Canons Burn?" *Anthropology Now* 12 (3):
34–38.

Traweek, Sharon, Duygu Kaşdoğan, and Kim Fortun. 2021. "Searching
for How Epistemic Power Is Made, Appropriated, Circulated, and

Challenged: An Interview with 2020 4S Bernal Prize Winner Sharon Traweek." *Engaging Science, Technology, and Society* 7 (2): 97–119.

United Nations. 2007. "United Nations Declaration on the Rights of Indigenous Peoples." http://www.un.org/esa/socdev/unpfii/documents /DRIPS_en.pdf.

Verhagen, Joachim. Science Humor. https://jcdverha.home.xs4all.nl /scijokes/index.html#2.

Vivian, Bradford. 2004. *Being Made Strange: Rhetoric beyond Representation.* Albany, NY: SUNY Press.

White, Hayden. 1992. "Writing in the Middle Voice." *Stanford Literature Review* 9: 179–87.

World Bank Organization. 2010. *Indigenous Knowledge for Development: A Framework for Action.* Washington, DC: World Bank Group. http:// documents.worldbank.org/curated/en/388381468741607213/Indigenous -knowledge-for-development-a-framework-for-action.

World Intellectual Property Organization (WIPO). 2024. *WIPO Treaty on Intellectual Property, Genetic Resources, and Associated Traditional Knowledge.* Geneva: World Intellectual Property Organization. https://www .wipo.int/edocs/mdocs/tk/en/gratk_dc/gratk_dc_7.pdf.

II Infrastructuring Postcolonialities

FIVE / LIV ØSTMO, JOHAN HENRIK M. BULJO,

LINE KALAK, AND JOHN LAW

Colonial Struggle and the Infrastructures of Knowing

A Story from Sápmi

In spring-winter (*giđasdálvi*), at the end of winter, Sámi people in Guovd-ageaidnu hunt ducks. In the past, fresh food was scarce on the subarc-tic plateau at this time of year, but they hunted just enough to give the family fresh meat. This hunt—it's called the *lodden*, the spring-winter duck hunt—is central to Sámi ecological and social knowledge, except that since the 1950s the tradition has been squeezed to near extinction. In recent decades the *lodden* has been only marginally legal, and its rules have been so restrictive that it has barely been practiced. For the Norwegian colonial state, spring-winter duck hunting has been environmentally dam-aging (though recreational autumn duck hunting, which is legal, is not). For Sámi people, recreational duck hunting at any time of year is unethical and offensive, and it is particularly damaging to shoot fledgling birds in the autumn.

This chapter is about colonial pressure, Sámi tactics of resistance, and two quite different ways of living in and knowing the world. About indi-geneity, colonialism, epistemic decentering, and political decentralizing in particular, it is also about resistance. Is it better to struggle within the rules and conventions and ways of knowing of a colonial state? Or to break those rules? Sámi experience suggests that this is a matter of tactics. Sometimes rule breaking works. Sometimes rule following is better. It all depends. But as many have noted, the dilemma is that to follow the rules of colonial states is also to reproduce the conditions of coloniality (Nadasdy 2003). And this is the particular focus of our chapter.

Our starting point is that all ways of knowing—colonial and otherwise—can be understood as practices that draw on but also help to reproduce heterogeneous resources or arrangements. Those resources are all the circumstances that make knowing possible at all, that generate what Michel Foucault (1976, xix) called "les conditions de possibilité," the conditions of possibility. And they are heterogeneous, because they are material and embodied, but also social, institutional, economic, linguistic, narrative, normative, methodological, epistemological, and metaphysical. This in turn implies that struggles about de/centering and de/centralizing are irretrievably tangled together. In this way of thinking, knowing in all its varieties is thus the art of drawing on and weaving particular more or less implicit heterogeneities together to make sense of and handle the world. But how to think about these heterogeneities?

In what follows, we talk of these as *infrastructures of knowing* (see Cardwell, Joks, and Law 2024). We use this phrase partly because these resources are more or less invisible: at least while they work, they are usually taken for granted. We also use the term because they shape but do not determine knowing practices. The latter are afforded by and depend on their infrastructural resources, but (as the analogy with utilities such as water or electricity suggests) the ways in which these are drawn on and woven together is at least partially underdetermined. Finally, we adopt the term because, as we said above, knowing practices also help to reproduce the infrastructures that are embedded in them. So there's a performative loop at work here because ways of knowing also help to make the conditions that give them life in the first place. We need, then, to think of knowledge and practices and their infrastructural resources in the same breath. All of these are done together.

This throws up various issues. One is that this performativity is often far from obvious for dominant ways of knowing. So, for instance, environmental biology or colonial administration are so widely distributed and performed that despite their periodic failures, neither they nor their ways of knowing disappear. Instead, any failure looks like a temporary glitch, and they continue to define reality. By contrast, the performativity of precarious knowing practices is very obvious. The *lodden* is a case in point. Its infrastructures are not being sustained elsewhere. A small part of the inland plateau aside, it has been squeezed out of existence everywhere in Sápmi. So if it is also choked by colonial government in its last surviving location, then the infrastructures that make it possible will disappear too. A whole tradition and a way of knowing and living on the land will simply vanish. Any possibility of epistemic decentering and political decentralizing will evaporate.

We explore this by considering the performative infrastructures of two quite different ways of knowing: the *lodden* itself on the one hand, and the work of a committee, the Loddenutvalget (the Lodden Committee) of the Guovdageaidnu Municipal Council on the other. This council is the local government for much of the subarctic, Sámi-speaking, Norwegian Finnmark plateau. Municipality, county, national government, the council is therefore a small cog in the Norwegian administrative machine. This means that it draws on and reproduces the material, administrative, political, institutional, and methodological infrastructures of that machine, and that these set limits to its conditions of possibility. That's the dilemma that we touched on above. But at the same time, though it is a part of the state apparatus, the municipality has also been battling more or less unsuccessfully for decades with other parts of the government machinery to defend the *lodden*. So, and to state the obvious, the council both lies within and is critical of that state and its ways of knowing. And the Lodden Committee Report is likewise caught in this performative dilemma. As it takes the fight to the enemy to argue for decentering and decentralizing, it also reproduces more or less repressive centering and centralizing state and science-related infrastructures. In what follows, we explore these different infrastructures of knowing and reflect on the tactics of resistance.

Four further introductory observations before we move on. First, colonial government and extraction in Sápmi reach back at least four centuries. It is a history of economic, political, linguistic, cultural, geographical, material, and environmental oppression. We cannot describe this here, but struggles about the *lodden* need to be read against the angry, wounding weight of this colonial history (Lehtola 2004 [1997]; Minde 2003). Second, when we talk of *Sámi* or *Norwegian* (*Norsk*) the distinction is real, but it is not binary. Instead, it indexes a complex and asymmetrical colonial entanglement. As a part of this, the division also tends to homogenize its two halves. But this is misleading, because in practice *Sámi* and *Norwegian* come in endlessly many different versions. Since our focus is on the colonialism of messy practices of knowing, we ask that the chapter not be misread as an exercise in identity politics. Third, three of the authors of the chapter are *árbečeahpit* (local knowledge bearers), Sámi activists, members of the Lodden Committee, and cosignatories of its report. In particular, Johan Henrik Buljo, who chaired the committee, has a lifetime's local environmental experience. We do not take this to be problematic, but readers need to understand that this chapter is both substantially self-ethnographic and reflects and extends a collective commitment to Sámi political activism. Finally, fourth, there is the issue of language. Originally drafted in English by the fourth author, John Law (a British academic in the field of science, technology, and society who has worked with Sámi colleagues for a decade), this chapter draws deeply on the research and reasoning of the Lodden Committee Report. That report—written in Norwegian— was mainly drafted by Line Kalak (an activist academic lawyer) and Liv Østmo (another activist, an academic anthropologist, and secretary to the committee). But the committee itself worked primarily in Sámi. Language is a part of the infrastructures of knowing, an issue to which we return below.

The *Lodden*

In winter it is cold in subarctic Sápmi. No buds, no leaves; there is snow on the ground and ice on the lakes and the rivers. Until all-weather roads came in the 1960s, fresh food might become scarce by the end of winter. In the eight-season Sámi year, the arrival of spring-winter, *giđasdálvi*, was eagerly awaited. People watched for its signs: cracks in the ice and the first sign of returning birds: swans, geese, and grazing ducks, and finally the diving ducks. This was—it is—the moment for the *lodden*. So Sámi people build blinds (*čilla*), and use decoy ducks (*čohkkánlottit*), and before shotguns or .22 rifles they trapped the ducks on small rafts (*boarri*), with snares (*giella*). The *lodden* was never a large hunt. As we mentioned above, it provided one or two meals for a family after a hard winter. And it went (it goes) with intensive knowledge of birds, their habits, and their environments, the product of generations of careful observation and adaptation by knowledge bearers (*árbečeahpit*). Skill-full practices. It is an example of traditional ecological knowledge (TEK), with its acute sensitivity to the unfolding patterns in the webs that make up what happens in a particular place, its own special embodied ways of seeing and listening, and its very particular practical skills. It is a profound way of knowing quite unlike academic biology, though it is rarely taken seriously by either biologists or bureaucrats.

Skills: some of these are physical. Building rafts. Mooring them in the right place. Making decoy ducks (in the past this was taught in schools). Locating blinds where the birds will gather (Loddenutvalget 2021, 28). Creating these blinds out of shrubs and trees or camouflage netting (traditional ways of knowing do not stand still). Handling guns in boats in ice floes with or without outboard motors. Pulling boats safely over ice. Knowing where to go to watch the ducks, sharing intelligence with other *árbečeahpit*, and in recent decades sharing intelligence about the movements of *garjját*, the "crows," the agents of the heavy-handed SNO, Statens Naturoppsyn (Norwegian Nature Inspectorate). Which birds to shoot and which not to shoot. How to shoot at close range so the bird dies instantly and does not suffer a lingering death. How to pluck and singe and clean and cook a duck. How to clean up afterward so that you leave no trace behind. And, that contemporary additional burden, how to do the ubiquitous paperwork that comes with every duck that is shot.

Physical skills like these go with those that are observational. We touched on some of these above. *Árbečeahpit* watch the land, the sky, the weather, the growing cracks in the frozen rivers and the lakes as they look for the returning ducks. They identify the species by how they look, fly, settle, swim, and dive. They know that female diving ducks stay underwater longer than males, distinguish the mating rituals and dances of different species (Loddenutvalget 2021, 30), observe how the posture of female ducks changes at mating time, and carefully watch the size of broods. And they listen too. "Each duck species has its own sound, and the names of ducks in Sámi describe the sound they make when they are flying [*jietna-dallá*]" (29). So, they also identify species from the faint sounds made by their wings as they fly far overhead.

A third overlapping set of skills is social. *Loddejeaddjit* talk about how this *gidasdálvi* compares with the last. What the arrival of the ducks this year might portend for future years. They work together to build blinds and rafts and decoy ducks. At the same time, young people learn by playing, hearing elders talk, watching what happens, and participating in the *lodden* (Loddenutvalget 2021, 32–33). But—this is important—the social skills of *árbečeahpit* also extend to interactions with animals including ducks. This is because, in the Sámi world, creatures are not objects but lively beings with wills and moral sensibilities worthy of respect in their own right (Oskal 2000; Magga, Oskal, and Sara 2002; Sara 2009). If you shoot a duck, then this is because you have communicated with it; you asked it to come, and it gave itself of its own volition (Loddenutvalget 2021, 44). As a part of this, you may also have *yoiked* it, sung for and to it, as a sign of respect. And this humility and respect does not stop with the death of the bird. You also eat it carefully, trying keep its skeleton intact rather than tearing it apart, and after the meal you show your gratitude by offering a blessing (*sivdnidit*; Sjöberg 2018) and protecting its skeleton by laying it under a tree.

Árbečeahpit, then, hunt care-fully (Loddenutvalget 2021, 16). They take a limited number of carefully chosen birds. No one hunts in the places where they nest. It is sacrilege to disturb nests because birds need peace, and you need to hold your breath if you get too close to a nest. So, for instance, *árbečeahpit* were horrified when ornithologists scared birds off their nests in a count by walking across a wetland with a cord stretching between them (37–38). And this respectful sociability is not confined to animals and birds but extends to (what outsiders take to be) natural, supernatural, or spiritual phenomena including lakes and rivers, the weather, and sacred places (*sieidi*) (Porsanger 2012). In short, the world for Sámi *árbečeahpit* is filled with lively and morally sensible beings, and the landscape (*meahcci*) is not a geographical area with features and populated by objects. Instead, it is a set of hopefully productive task-related social encounters with other ethically conscious beings. As Audhild Schanche (2004, 169) notes, at least in the past people negotiated with *meahcci*, not about it.

These physical, observational, and social skills also embody a sense of contingency (Loddenutvalget 2021, 44). This is a world of processes and relations rather than of objects in an environment (Mazzullo and Ingold 2008). The calendar and the clock have no relevance for the *lodden*. People say, "*Beaivvit eai leat badjálagaid, muhto maŋŋálagaid*" (the days are not on top of one another, they are one after the other). You know you cannot follow a fixed schedule. This stress on events and happenings is reflected in the Sámi language, which easily makes verbs out of nouns, and vice versa, and often attends to processes rather than designating attributes. It is reflected, too, in what Sámi call *meroštallan*: the process of making judgments by watching, observing, and relating different features of the environment together. Here things don't get fixed or counted. Instead, *meroštallan* is about sensing the shifting web of relations between the endlessly variable encounters that make up different (plural) *meahcit*. So, Sámi *árbečeahpit* do not imagine that they can fully know and control the world. Instead, they interact with it by attending to and responding to it as it unfolds, knowing that life is uncertain and plans often come unstuck. A final point: *árbečeahpit* are also resourceful and adaptable and know how to cope with the unexpected (Loddenutvalget 2021, 35). There is a word, *birget*, that catches what is at stake. Roughly this means skillfully and flexibly sustaining a livelihood and a good life in changing circumstances—and doing so in ways that balance concerns that (in outsiders' language) are simultaneously economic, social, spiritual, individual, and environmental. "Ethics and spirituality are intertwined," writes the Lodden Committee. "They offer guidelines for who you are in the world you live in and how you should relate to your surroundings and to all the beings, or actors, that surround you" (44).

Interlude 1

The *lodden* is a set of practices and a way of knowing that grows out of, depends on, and sustains heterogenous resources that together make up its own infrastructures. What are the features of those resources?

The stories above tell us that these practices are locally done in particular observational and practical ways with specific tools and in particular social relations. *Árbečeahpit* are authorities, know how to observe ducks in their contexts, and know how to talk with one another. They carry physical and cognitive skills (think of *birget*), the practical and more or less modest subjectivities that these demand, and a specialist vocabulary. They have narratives too. Shared in talk and song, these narratives describe a world of contingent encounters where creatures and other features of the environment are social and ethical, and demand respect and reciprocity. In saying that the hunt is conducted to feed the family, we've also implied something about economics: relations of respect that are neither commercial nor recreational. We've touched, again implicitly, on its epistemological assumptions: this is a world that cannot be fully known or counted. Knowing, for instance through *meroštallan*, is modest. But repeating interactions between (often social) beings can often be understood in particular places by those with appropriate experience. And finally, as a part of this, we've also hinted at its metaphysics. On the one hand, the realities of this world are relational, social, ethical, and ultimately uncertain; and on the other hand, there is considerable continuity between what outsiders might think of as nature on the one hand, and culture on the other.

Such are some of the infrastructural resources—social, practical, epistemic, and metaphysical—that the practices of the *lodden* draw on and weave together; that set its conditions of possibility and allow it to reproduce itself. People become *árbečeahpit* by practicing the *lodden*, by becoming *loddejeaddjit*. And children learn about it informally by playing, watching, listening, and moving from small tasks to take on larger responsibilities. Except, as we have also said, the practices that sustain these infrastructures are under threat. The *lodden* is being remorselessly squeezed, and the learning that is needed to sustain its infrastructures is critically at risk.

The Lodden Committee

The Lodden Committee Report takes us to a very different way of knowing that draws on and weaves together a more or less different set of infrastructures. More than a hundred pages long and extensively researched, it describes the *lodden*, the substantially successful attempts by the state to restrict it since the 1950s, and the local struggles to resist that pressure. The report was commissioned by the Guovdageaidnu Council in 2020 in response both to upcoming national and Sámi elections, and to a national review of the long-term future of the *lodden*. It is a tactical epistemic and political intervention pressing the case for both decentering and decentralizing. It argues for the legality of the *lodden*, greater Sámi control over its conduct, its value as an ecological and cultural indigenous practice, and for the rights of the Sámi as an indigenous people to sustain their culture and practices. It situates its arguments in local ecological knowledge, but also in literatures drawn from national and international law, social anthropology, and biology. In short, it takes the fight to the enemy. But how does it do this? What are its tactics? What are the infrastructures that it draws on as it weaves its intervention?

A first response to these questions takes us to language. We said this earlier. The committee talks and gathers local stories in the Sámi language, but its paperwork is primarily in Norwegian. Indeed, the mayor commissioned the report in Norwegian precisely because it is targeted at powerful outsiders who neither understand nor read Sámi. In one way, this was not an inconvenience: having been subjected to a colonial education system, the members of the committee are bilingual. But it also means that the Sámi language is sidelined, and the result is that many Sámi realities are subordinated. So, for instance, sometimes the report needs to deal with straightforward mistranslations. The Norwegians call the *lodden* the spring hunt (*vårjakt*), but it isn't. As we saw above, it's a spring-winter hunt, which means that it is over before the ducks start to breed. The report also needs to point to more subtle mismatches between Norwegian and Sámi words and the practices and realities that they index. As we again saw earlier, Sámi people talk of *meahcit*, task-related places and relations (Joks, Østmo, and Law 2020, 307; Ingold 1993). However, there is no equivalent term in Norwegian, and in Norway *meahcci* gets translated as *utmark*. As a result, the report needs to explain that this is a deeply consequential mistranslation because *utmark* is an agricultural term (it denotes land lying beyond the cultivated fields of a farm used, for instance, for rough summer grazing) that has nothing to do with the relations and encounters of *meahcci*. And—another infrastructural inconvenience that is very difficult to handle—this shift from processes to objects extends into differences between the ways the two languages work. This is subtle, but much more than in Norwegian (or English), the Sámi language stresses processes and relations rather than objects. As we mentioned above, in Sámi it is relatively easy to make verbs out of nouns and vice versa. So, bilingual though they are, Norwegian was nevertheless an awkward tool for the members of the committee. Infrastructural realities woven into Sámi practices that are easily expressed in the Sámi language became counterintuitive in Norwegian language and practice, and for this reason many of these are laboriously spelled out in the report.

Language and practice cannot be pried apart. The world of the Sámi, their environmental knowledge, the way in which they live, their material practices, and how they talk are all woven together. All are part of the infrastructures of knowing. So what can we say of the material practices of the committee? To think about this well, we need to attend to infrastructures that are so mundane—even tedious—for office workers that they usually escape attention. For instance, when the committee met, its members sat around a table on blue chairs. They had pens, notepads, piles of paperwork, mobile phones, computers, and a computer projector. And they met in a room located in the Guovdageaidnu Council offices. All of these are unremarkable mundanities—so much so, indeed, that they are barely noticeable. But they are also important because they remind us that we are no longer outdoors in the world of the *lodden*. Instead, we are in the materially very different world of the office. These are the kinds of infrastructural resources that make office work and office ways of knowing possible. And it is a world much more closely linked to the practices of academics or Oslo administrators than to those of the *lodden*.

Comments on this: office work is substantially textual. So committee members worked with internet files, legal and policy documents, historical accounts of the *lodden*, ethnographies of other circumpolar indigenous peoples, and academic papers about bird populations from environmental journals. Some of its members visited archives to look through old, undigitized documents. Then they assembled those documents, took notes, made summaries, cut and pasted paragraphs, and wrote drafts to make the report. So, yes, like the *árbečeahpit*, they talked. But unlike the *árbečeahpit*, they also traded in the texts central to the infrastructures of office work and its ways of knowing. Like any other committee in the world of politics or administration, they drew on written resources, they wove these together, and what they knew took the form of—another text, a report that fitted the knowledge practices and infrastructures of government, a way of knowing legible to the state. "Dáža lea hárjánan dasa ahte juohke ášši galgá leat čáhppat vielgada alde ovdal go sáhttá mearridit" (The Norwegians need to see the case, black on white, before making any decision). This is what Sámi people say when they think about the state. To take the fight to the enemy, they needed to practice knowing like a state. They needed to draw on state infrastructures. And they needed to weave these together in state-relevant ways.

"Black on white." But what kind of black on white? Policy-relevant texts are simultaneously representational and administratively normative. They define a problem; describe the parts of the world relevant to that problem; draw inferences from those descriptions; counter alternative possible inferences; and arrive at conclusions that resolve the problem that they began with. Of course, we are not naive. There is lobbying in the corridors of power, and there are endless hidden agendas. But texts are infrastructural resources crucial to this administrative and political game. They may or may not change anyone's mind, but they are essential because they and the arguments they carry are needed to justify the political and administrative decisions that are actually taken. To put it succinctly, administration trades in justifications (Boltanski and Thévenot 2006). And that is how you approach policy-relevant documents. You write them as justifications.

How to do this? The material answer is that your text is probably linear. It's a narrative that runs nice and smoothly, A, B, C. It's probably hierarchical too, talking up the most important parts of the narrative and talking down less important stories. And then it really needs to loop back on itself to close its narrative arc. Problem, loop, problem solved—that's the formula for policy justification. And such is the logic that structures the report. Title, authorship, foreword, and summary. Here it puts hierarchy to work. Everything the busy reader needs to know is already there at the top of the tree. And then there are the chapter titles that point to its major narrative steps (Loddenutvalget 2021). Here they are translated into English:

At the end (here's the looping back) it solves its introductory problem in the final "Way Forward" section. Meanwhile (this is hierarchy again), substeps in the argument get downgraded, as, for instance, in the subheads for chapter 11 (Loddenutvalget 2021):

11 What Threatens Ducks and Geese?
    11.1   Autumn Hunting
    11.2   Pollution
    11.3   Development
    11.4   Predators
    11.5   Fishing Boats/Trawlers

Or substeps may be demoted even further to footnotes and references. So the implicit message is that most readers don't need to attend to such details. But there is a second crucial message too. This is that the report is not simply drawing on local knowledge but is justifying its arguments by drawing on external—and therefore administratively reliable—authorities.

Which authorities? The answer is: three in particular. First, it draws on anthropology to say that Norwegian and Sámi cultures are different (Loddenutvalget 2021, 83); that Norwegian culture is dominant; that Sámi daily life depends on the recognition of this difference; that TEK is both a form of culture and a way of life (Loddenutvalget 2021, 11, 17); that the law reflects Norwegian culture, with disastrous consequences for the *lodden*; and that spring-winter duck hunting is practiced by circumpolar indigenous peoples and accepted by governments in Fennoscandia, Russia, the United States, and Canada.

Second, it draws on the law, a move that also demands a fair bit of anthropological (mis)translation. Culture (Norwegian *kultur*)? This is a term foreign to both the Sámi and to contemporary anthropology. But there it is in the report anyway, and this is because the law is a form of expertise, an authority, and relevant parts of it are written in ways that require the protection of "cultures." Which means in turn that the report needs to cite *culture* if it is to use national and international law as an infrastructural resource—which is what it does. So it says that the *lodden* is consistent with the laws on biodiversity and indigeneity; that there is a legal requirement to draw not just on scientific knowledge but also on local traditional knowledge in the management of natural resources; that the *lodden* is protected because minority indigenous cultures, identities, and practices are legally protected; that the Sámi have the legal right to participate in the management and conservation of natural resources, to practice their culture, and pursue traditional activities that ensure cultural and economic self-preservation; and that the conditions needed to transmit the *lodden* and its knowledges are similarly subject to legal protection. Without *culture*, then, the law is no longer an infrastructural resource.

Third, the report also draws on environmental and biological science (Loddenutvalget 2021, 49–54) because its authors correctly assume that administrators and policymakers take science to be more or less reliable. And here it says, first, that given its scope and its careful conduct, the *lodden* threatens neither duck populations nor biodiversity. Then, second, it describes what it takes to be the real threats to duck populations (citations included). These include the Norwegian recreational autumn duck hunt. This has no quotas; female ducks are targeted; vulnerable fledglings are killed or scared off; and protected species are at risk because autumn plumage makes it difficult to identify ducks. The report then goes on to describe the effects of oil and plastic pollution on seabirds and argues that wind turbines, masts, buildings, and power lines kill or affect birds and their migration patterns; that protected predators have detrimental effects on bird populations; and that there are probably substantial bird losses to fishing boats and trawlers.

Interlude 2

Like the *lodden* itself, the Lodden Committee Report is a way of knowing that draws on, weaves together, and reproduces a particular heterogenous infrastructure. Yes, it draws on the knowledge of *árbečeahpit*, but most of its infrastructural resources—and the ways in which it uses these—instead resonate with the state infrastructures of knowing.

Aspects of this. Linguistically it shifts to, makes use of, and reinforces a colonial Norwegian language infrastructure. Institutionally it draws on professional authorities and their methodological, argumentative, and theoretical tools together with the specialist academic divisions of labor and ways of knowing that are taught, credentialed, and practiced in formal institutions. Not coincidentally, the committee also works textually because its narratives need to travel to the centers of power. This means that it works materially and practically in broadly the same way as any government or university office. At the same time, the resulting textuality remolds what it is to know (Ong 1998). The consequences of this depend on circumstances: texts and forms of writing may be subversive (Cole 2002; Smith 2012). However, to write in a state-legible way, as the report does, is to separate descriptions from the realities they depict. As a part of this, it is to weave a range of representations together to make a single depiction in the form of an epistemically centralized overview that is arguably also organizationally centered (Haraway 1988; Latour 1990). Thus this hierarchical literary form is reproduced—indeed mimicked—in the institutional infrastructures of knowing in (and beyond) Norway, which distinguish centers from peripheries in both science and policy. It also, however, opens up novel methodological and epistemological issues. Are our descriptions good enough? And how can we demonstrate this? These are the kinds of questions that arise—and it is no coincidence that much of the textual structure of the report weaves together resources (for instance, drawn from anthropology and the law) in ways that are intended to handle these questions. Finally, these textual, methodological, and epistemological infrastructures in turn reproduce implicit metaphysical assumptions. This is because the very business of creating overviews and moving descriptions from place to place presupposes the existence of a single reality out there endowed with a definite and more or less discoverable form. It also assumes that the basic mechanisms that shape reality are everywhere similar (Law 2004). And, at least in the context of the report, it also assumes that reality is binary in at least two ways. First, it takes it for granted that nature, with its relations of causality, is quite unlike culture, with its normatively charged relations between lively sentient beings. Second, it also assumes that facts about nature are distinct from the values and the decisions that likewise reside within that culture.

Conclusion

In 1910, herder, hunter, and author Johan Turi published the first book of Sámi literature. Writing of reindeer herding, hunting, trapping, healing, and shamanism, in the preface he observed that "the Swedish government wants to help us as much as it can, but they don't get things right about our lives and conditions." And this, he implied, is partly because there is a link between thinking and material circumstances: "When a Sámi becomes closed up in a room, then he does not understand much of anything. . . . His thoughts don't flow because there are walls and his mind is closed in. . . . But when a Sámi is on the high mountains, then he has quite a clear mind" (Turi 2012 [1910], 11). Perhaps, he thought, if they were given texts, then the bureaucrats and politicians might understand Sámi people better. Hence his book: "Herein [this book] are all sorts of stories, but it isn't certain whether they are true, since they haven't been written down before" (13).

The Lodden Committee Report follows Turi. It resists state ways of knowing and ordering by using state-relevant representational and organizational infrastructures far removed from those of the Sámi knowledge bearers. But this also creates a painful dilemma. Yes, it turns those infrastructures against themselves and makes the *lodden* potentially legible to state power (Scott 1998). But as we have seen, at the same time it reproduces the centered and centralized infrastructures of government and natural science (Nadasdy 2003).

Table 5.1 is too simple, but it both summarizes the argument of this chapter and underscores this dilemma. Thus it reveals both the heterogeneity of the infrastructures of knowing and the way in which epistemic decentering and political decentralizing are tangled together. So what to make of this?

Table 5.1 Two Infrastructures of Knowing

Infrastructures	*Lodden*	Report (state/science)
Material forms	**Embodied knowing:** observation, sensitivity to context, talk, embodied, practical; socially decentralized	**Textual knowing:** also embodied; crafting, organizing, and juxtaposing mobile texts; susceptible to epistemological centering and political centralization
Social/ institutional forms	**Local and distributed** between experienced authorities, *árbečeahpit* and *loddejeaddjit*; socially decentralized	**Specialist, hierarchical,** and **centered,** with division of labor between different authorities
Learning	**In practice:** from local authorities, *árbečeahpit*; participatory; generational; decentralized	**Formal training:** specialist expertise, credentialed; centralized; probably centered
Subjectivities	**Modesty;** knowing is only local; decentralized, possibly decentered	**Referential mastery;** fits with centered and centralized narratives (see below)
Language	**Sámi,** verbs; fits with relational narratives (see below)	**Norwegian,** nouns; resonates with authoritative centralized descriptions
Economic forms	**Relations of respect** in a context of necessity	More likely to be **legal** and (quasi) **contractual**
Narratives	**Distributed and relational;** song; talk	**Centered,** linear-hierarchical loops, means-ends; often textual
Interventions	Contingent and **uncertain,** *birget*	The aspiration is to centralized practical **control**
Judgments	**Unfold** in practice, shifting; *meroštallan*; distributed, decentralized	Formal and centralized **decisions** based on gathered evidence
Methods for knowing	**Practical,** qualitative, and depend on experience	**Formal,** in principle standardized, distanced, detachable, centralized, and centered; sometimes quantitative

Ethics	The world is an **ethical web;** personal ethics respect that web and its inhabitants	The **world is ethically neutral,** a set of causal relations; ethics belongs to the social and political
Epistemologies	**Practical,** situated, local, place-bound, modest, and about detecting limited forms of patterning; relatively decentered	**Referential** or representational, about gaining accurate summary overviews; good ways of knowing are transportable; therefore centered and centralized
Metaphysics: realities are	**Relational,** a shifting web of ultimately uncertain ethically extended social and reciprocal relations	**Determinate and binary;** nature has a definite form shaped by unchanging underpinning causal relations; the social world is made of interactions between ethically sensible, meaning-carrying actors
Metaphysics: spaces are	**Encounters,** negotiations; therefore decentralized	Events, objects, and causal relations in a **space-time box**

One thought is that the best form of resistance depends on circumstances. Indeed, this is a lesson that we draw much more widely from the history of Sámi resistance to Norwegian colonialism. These tactics have been hugely varied. So people have often simply ignored the state and carried on hunting or fishing illegally. They have used the classic tactics of passive resistance, including foot dragging, apparent compliance, passivity, "laziness," and "misunderstanding" (Lehtola 2018, 30). They have taken direct, extralegal action—most notably against the Alta hydroelectric dam (Briggs 2006), and by occupying Tiirasaari Island in protest against Deatnu salmon fishing restrictions (Uutiset 2017). They have created artworks (Hætta 2022), for instance, to protest against restrictions on reindeer herding with an installation outside the Norwegian parliament (Associated Press 2017). They have gone to law, for example, about fishing (Kalak and Johansen 2020), to resist mining development (Broderstad 2015, 19), and they won a spectacular victory in 2021 in the Norwegian Supreme Court when it ruled that the rights of Sámi herders had been ignored in the siting and construction of a major wind farm (Agence France-Presse 2021; Hess 2023). They have undertaken long-term constitutional engagement and have achieved limited political decentralization (Falch, Selle, and Strømsnes 2016). They have worked long and hard to explore how schooling and educational curricula might transmit traditional skills and ways of knowing to young people who are no longer raised on the land (Guttorm 2011). And they have authored academic and policy interventions in many contexts, including reindeer herding (Benjaminsen et al. 2015), salmon fishing (Joks 2015), powan fishing (Østmo and Law 2018), and land use (Schanche 2004; Joks, Østmo, and Law 2020).

The lesson we draw, then, is that how best to engage with powerful ways of knowing and the centralizing institutions in which these are embedded is a matter of tactics; that there are—that there can be—no hard and fast rules about this. Though, and as a part of this, we have a further thought. This is that it is often (usually?) wise to resist the temptation to treat the state and its infrastructures (including those of natural science) as if they were monoliths. Yes, they are powerful. Yes, they squeeze indigenous practices, their ways of knowing, and their infrastructures. And yes, they certainly feel like monoliths. (There are far too many struggles in Sápmi that parallel those of the *lodden*.) But if we imagine and interact with these powerful infrastructures by treating them as if they were solid all the way through, this too is performative. It probably helps to make them bigger. Instead, our suggestion is that it is wise to hold onto the idea that powerful institutions and infrastructures are neither consistent nor coherent—and then to look for ways of making use of those inconsistencies. This is what Johan Turi did over a century ago. He used the tools of textuality against those who governed. And as we have seen, so too does the Lodden Committee. Rather than treating power as a monolith, it levers the gaps and the differences between different versions of power and different ways of knowing, between the law, environmental science, anthropology, and the practices of policing. It borrows and plays with hierarchical texts and modes of storytelling, formal authorities, centralized institutional and political structures, centered epistemological assumptions, and binary nature/culture metaphysics. And it does this in the hope of loosening their grip. State-friendly infrastructures these may be, but they are also tools for undoing epistemic centering and political centralization.

These, then, are our main conclusions: that what we have called the infrastructures of knowing are heterogeneous; that centering, decentering, centralizing, and decentralizing are tangled together; that effective political resistance to centering and centralizing is a matter of tactics; and that it is wise to resist the assumption that power is monolithic. And a further political and methodological observation: to resist well, it also helps to attend to the performativity of knowing practices, to attend to what they are doing. And—this is the core point—to attend to what they are doing by stealth, to the implicit social, institutional, epistemological, and meta-physical infrastructures that they are reenacting. This is a political impera-tive. It is really important to know what texts, policy reports, or scientific claims are doing in the places where they circulate, doing not only overtly but also by implication, doing without anyone knowing (Law 2011), their hidden agendas. But this implies a further methodological correlate. To do this well—to attend to what is implicit—we also need to burrow into the specificities of infrastructural practices, to pick through what happens moment by moment even if (indeed precisely because) this is so mundane. And this is because, to use the English-language idiom, the devil is in the detail, and what that detail is doing. And this is what we have attempted in this chapter on the *lodden*. As we have looked at the work of the Lodden Committee we have attended to its language politics; to its unremarkable form as a meeting of people sitting around a table in a room; to its searches in the archives, digital and otherwise; to the way it drafted and juxtaposed its textual bits and pieces; to the hierarchical and centering character of its report; to how it drew on authorities from outside Sápmi; and to its state-friendly methodological, epistemological, and metaphysical performativ-ity as it took the fight to the enemy. It is by picking through the detail that we begin to discover the performative depths of the practices of knowing, that we learn how power enacts itself—and that there are no shortcuts.

We intend no criticism of the committee or its work. It made the tacti-cal infrastructural compromises needed to intervene in the politics of the *lodden*. It centered and centralized in an attempt to resist centering and centralizing. It did what it did to try to secure a future for the *lodden*. But as we have also said, there are many other possible forms of resistance. So, the issue—again always tactical—is how to rearticulate ways of knowing by weaving together infrastructures that also keep indigenous traditions alive. "Jahki ii leat jagi viellja" (One year is not the next year's brother). Sámi people have always adapted to circumstances that they know they cannot control, and they continue this tradition in the struggles of the twenty-first century.

Afterword: What Happened Next

As we completed the first draft of this chapter, the Guovdageaidnu munici-pality was in the process of submitting the Lodden Committee Report to Oslo. After nine months of silence in which the ministry digested the various

submissions it received, the state secretary in the Klima- og Miljødepartement (Ministry of Climate and the Environment) announced an unprecedented visit to Guovdageaidnu to consult with local stakeholders. He
came, he met and talked with local people, and he went away again to
consult with other stakeholders including the Sámi parliament. Some
locals thought he might be impressed by the depth of knowledge of the
loddejeaddjit and the care with which the hunt was practiced. Then there
was a further period of silence, but just days before we submitted the
final version of this chapter, the ministry published its decision. Extending permission for the *lodden* for five more years, the preamble to its
ruling cited the importance of the practice for Sámi tradition, culture,
and customary rights and the need for "ecological and cultural sustainability" (Lovdata 2023.) It agreed that more ducks (and more appropriate diving ducks) could be taken. And female ducks too. The length of
the hunt (though not its area) was slightly increased and made more
subject to local control. And, though this is not in the protocol, a five-
year study of the consequences of the *lodden* on duck numbers is also
being discussed.

For the *loddejeaddjit*, this was a substantial victory. (Remember that
many thought that the *lodden* would simply be forbidden.) But it was still
only a partial victory. The new policy was only partly decentralized (quotas, the permitted period for the hunt, and the total number of days within
those dates, together with hunting locations, were still being stipulated
from the center). It was only partially decentered (some, but only some,
Sámi TEK had indeed informed the decision). And the ministry ruling
didn't mention the national and international law that in theory protects
Sámi culture and practice in Norway. Even so, something had changed.
But why? The answer is that we do not know, but two possibilities seem
likely. First, there was a major political and legal conflict raging in and
beyond Norway about a wind farm controversy at Fosen that we briefly
mentioned above. Those who were thinking about the *lodden* in Oslo did
not, we guess, consider that this was a good moment to provoke a further
conflict about Sámi rights. Second, it also seems possible, even likely, that
writing a report that made the *lodden* legally, scientifically, and especially
anthropologically legible to Oslo politicians and civil servants was important too. It reminded the center that the law takes Sámi culture and practice
seriously. And it also revealed the depth and subtlety of local environmental knowledge. Perhaps—again, we do not know—it was the latter that led
to the ministerial visit to the North.

Clearly, this remains political work in progress: the struggle for the *lodden* will go on. And indeed, the next step in that struggle is clear. It will be to try to persuade the ministry that the five-year environmental study of ducks that is under discussion should rest not just on biological science but also draw on the expertise of the *árbečeahpit*, the Sámi knowledge holders. The next struggle, then, will be a struggle to decenter. It will be about finding a place for TEK, its methods, and its practices in future environmental research in Guovdageaidnu. We keep our fingers crossed.

References

Agence France-Presse. 2021. "Norway Court Rules Two Windfarms Harming Sami Reindeer Herders." *Guardian*, October 11, 2021. https://www .theguardian.com/world/2021/oct/11/norway-court-rules-two-windfarms -harming-sami-reindeer-herders-turbines-torn-down.

Associated Press. 2017. "Sami Artist Protests Norwegian Reindeer Cull with Unique Public Art Piece." CBC, Radio Canada, December 7, 2017. https://www.cbc.ca/news/canada/north/sami-reindeers-norway-1 .4437517.

Benjaminsen, Tor A., Hugo Reinert, Espen Sjaastad, and Mikkel Nils Sara. 2015. "Misreading the Arctic Landscape: A Political Ecology of Reindeer, Carrying Capacities, and Overstocking in Finnmark, Norway." *Norsk Geografisk Tidsskrift—Norwegian Journal of Geography* 69 (4): 219–29. https://doi.org/10.1080/00291951.2015.1031274.

Boltanski, Luc, and Laurent Thévenot. 2006. *On Justification: Economies of Worth*. Princeton, NJ: Princeton University Press.

Briggs, Chad M. 2006. "Science, Local Knowledge and Exclusionary Practices: Lessons from the Alta Dam Case." *Norsk Geografisk Tidsskrift— Norwegian Journal of Geography* 60 (2): 149–60. https://doi.org/10.1080 /00291950600723146.

Broderstad, Else Grete. 2015. "The Finnmark Estate: Dilution of Indigenous Rights or a Robust Compromise?" *Northern Review* 39: 8–21.

Cardwell, Emma, Solveig Joks, and John Law. 2024. "The Infrastructures of Knowing." In *Elgar Encyclopedia of Science and Technology Studies*, edited by Ulrike Felt and Alan Irwin, 43–51. Cheltenham, UK: Edward Elgar.

Cole, Peter. 2002. "Aboriginalizing Methodology: Considering the Canoe." *International Journal of Qualitative Studies in Education* 15 (4): 447–59. https://doi.org/10.1080/09518390210145516.

Falch, Torvald, Per Selle, and Kristin Strømsnes. 2016. "The Sámi: 25 Years of Indigenous Authority in Norway." *Ethnopolitics* 15 (1): 125–43. https://doi .org/10.1080/17449057.2015.1101846.

Foucault, Michel. 1976. *The Birth of the Clinic: An Archaeology of Medical Perception*. London: Tavistock.

Guttorm, Gunvor. 2011. "Árbediehtu (Sami Traditional Knowledge)—as a Concept and in Practice." In *Working with Traditional Knowledge: Communities, Institutions, Information Systems, Law and Ethics*, edited by Jelena Porsanger and Gunvor Guttorm, 59–76. Guovdageaidnu/ Kautokeino: Sámi allaskuvla/Sámi University College.

Haraway, Donna J. 1988. "Situated Knowledges: The Science Question in Feminism and the Privilege of Partial Perspective." *Feminist Studies* 14 (3): 575–99.

Hætta, Susanne. 2022. "The Crows Take Everything: Interview with Britta Marakatt-Labba 11.04.22." *Kunstkrittik/Nordic Art Review*. https:// kunstkritikk.com/the-crows-take-everything/.

Hess, Martine Aamodt. 2023. "Norway's Treatment of Sámi Indigenous People Makes a Mockery of Its Progressive Image." *Jacobin*, March 13, 2023. https://jacobin.com/2023/03/norway-sami-indigenous-people -reindeer-herding-wind-turbines-dispossession-protest.

Ingold, Tim. 1993. "The Temporality of the Landscape." *World Archaeology* 25: 152–74.

Joks, Solveig. 2015. "'Laksen trenger ro': Tilnærming til tradisjonelle kunnskaper gjennom praksiser, begreper og fortellinger fra Sirbmá-området" ["The salmon needs peace": Approaching traditional knowledge through practices, concepts and stories from the Sirbmá area]. PhD diss., Tromsø UiT, Norges Arktiske Universitet.

Joks, Solveig, Liv Østmo, and John Law. 2020. "Verbing *meahcci*: Living Sámi Lands." In "On Other Terms: Interfering in Social Science English," edited by Annemarie Mol and John Law, special issue, *Sociological Review* 68 (2): 305–21.

Kalak, Line Aimee, and Bjarne Johansen. 2020. *Tradisjonell kunnskap og forvaltning av sjølaksefisket, Dieðut*. Guovdageaidnu: Sámi Allaskuvla.

Latour, Bruno. 1990. "Drawing Things Together." In *Representation in Scientific Practice*, edited by Michael Lynch and Steve Woolgar, 19–68. Cambridge, MA: MIT Press.

Law, John. 2004. *After Method: Mess in Social Science Research*. London: Routledge.

Law, John. 2011. "Collateral Realities." In *The Politics of Knowledge*, edited by Fernando Domínguez Rubio and Patrick Baert, 156–178. London: Routledge.

Lehtola, Veli-Pekka. 2004 [1997]. *The Sámi People: Traditions in Transition*. Fairbanks: University of Alaska Press.

Lehtola, Veli-Pekka. 2018. "Evasive Strategies of Defiance—Everyday Resistance Histories among the Sámi." In *Knowing from the Indigenous North: Sámi Approaches to History, Politics and Belonging*, edited by Thomas Hylland Eriksen, Sanna Valkonen, and Jarno Valkonen, 29–46. Abingdon: Routledge.

Loddenutvalget. 2021. *Lodden—en samisk kulturbærende sedvane: En utredning om lodden i Guovdageaidnu*. Guovdageaidnu: Guovdageainnu suohkan/Kautokeino kommune.

Lovdata. 2023. "Forskrift om lodden (vårjakt på ender) fra og med 1. mai 2023–6. juni 2028, Kautokeino kommune, Troms og Finnmark" [Regulation on the lodden (spring duck hunting) from and including 1 May 2023–6 June 2028, Kautokeino municipality, Troms and Finnmark]. Lovdata, accessed April 29. https://lovdata.no/dokument/LF/forskrift /2023-04-04-483.

Magga, Ole Henrik, Nils Oskal, and Mikkel Nils Sara. 2002. *Animal Welfare in Sámi Culture.* Guovdageaidnu: Sámi Allaskuvla.

Mazzullo, Nuccio, and Tim Ingold. 2008. "Being Along: Place, Time and Movement among Sámi People." In *Mobility and Place: Enacting Northern European Peripheries*, edited by Jørgen Ole Bærenholdt and Brynhild Granås, 27–38. Aldershot: Ashgate.

Minde, Henry. 2003. "Assimilation of the Sami—Implementation and Consequences." *Acta Borealia* 20 (2): 121–46. https://doi.org/10.1080 /08003830310002877.

Nadasdy, Paul. 2003. *Hunters and Bureaucrats: Power, Knowledge and Aboriginal-State Relations in the Southwest Yukon.* Vancouver: UBC Press.

Ong, Walter J. 1988. *Orality and Literacy: The Technologizing of the Word.* London: Routledge.

Oskal, Nils. 2000. "On Nature and Reindeer Luck." *Rangfer* 20 (2–3): 175–80.

Østmo, Liv, and John Law. 2018. "Mis/translation, Colonialism and Environmental Conflict." *Environmental Humanities* 10 (2): 349–69.

Porsanger, Jelena. 2012. "Indigenous Sámi Religion: General Considerations about Relationships." In *The Diversity of Sacred Lands in Europe: Proceedings of the Third Workshop of the Delos Initiative*, edited by Jose-Maria Mallarach, Thymio Papayannis, and Rauno Väisänen, 37–45. Gland, Switzerland: International Union for Conservation of Nature.

Sara, Mikkel Nils. 2009. "Siida and Traditional Sámi Reindeer Herding Knowledge." *Northern Review* 30: 153–78.

Schanche, Audhild. 2004. "Horizontal and Vertical Perceptions of Saami Landscapes." In *Landscape, Law and Customary Rights*, edited by Michael Jones and Audhild Schanche, 1–10. Guovdageaidnu: Sámi Instituhtta.

Scott, James C. 1998. *Seeing Like a State: How Certain Schemes to Improve the Human Condition Have Failed.* New Haven, CT: Yale University Press.

Sjöberg, Lovisa Mienna. 2018. "Att Leava i Ständig Välsignele: En Studie av Sivdnidit som Religiös Praxis." PhD diss., University of Oslo.

Smith, Linda Tuhiwai. 2012. *Decolonizing Methodologies: Research and Indigenous Peoples.* 2nd ed. London: Zed.

Turi, Johan Mathis. 2012 [1910]. *An Account of the Sámi* [Muitalus sámiid birra]. Kárášjohka: ČálliidLágádus.

Uutiset. 2017. "Sámi Activists Occupy Island in Protest at Tenojoki Fishing Rules." *Yle*, July 17, 2017. https://yle.fi/uutiset/osasto/news/sami _activists_occupy_island_in_protest_at_tenojoki_fishing_rules /9717663.

SIX / ANGELA OKUNE, DUYGU KAŞDOĞAN, AALOK
KHANDEKAR, MAKA SUAREZ, AND KIM FORTUN

Remooring Academia

Postcolonial and Infrastructural Challenges

"Unfortunately, I don't circulate unpublished work for citation!" This was the response received from the host of an event series focused on decolonizing the university when emailed asking if his opening presentation was available to cite. He had beautifully presented many of the issues regarding post-structural adjustment changes to African university systems that Angela Okune wanted to build on by citing his work and drawing others to it. But the request was refused because the presentation hadn't yet been published (in a peer-reviewed academic journal, it was implied). The secrecy and proprietary sensibility encountered through this event and others has been surprising to us, especially coming from scholars explicit about their commitments to decolonizing knowledge production. Even though these events largely occurred at public universities, presentations were rarely open to outsiders, recordings of the events weren't available, and there were never any meeting notes or reflective blog posts. The events were enclosed, on many different registers (also see chapter 5, this volume).[1]

The enclosures of critical scholarship on decolonizing the university are contradictory, and all the more so when done within the privileged spaces of elite universities, some well known for supporting open science and open-access publishing, as well as diversity and "inclusive excellence." In part, this can be explained by what could be called the capture of open science and open access by elites, such that elite scholarship has become more widely available in recent years but through mechanisms that further marginalize scholars at less elite institutions, particularly those in the Global South (Knöchelmann 2019, 2021; Okune et al. 2021; Posada and Chen 2017;

Nkoudou 2020; Piron 2018; Sariola 2021). Elites are able to claim openness without structural, transversal change. The contradiction also has other threads, operating at a more personal level. In this chapter, following Sharon Traweek (chapter 4, this volume; Traweek, Kaşdoğan, and Fortun 2021), we study practices through which colonial epistemic authority is reproduced despite efforts to decenter knowledge making. Scholarship on the coloniality of academic research has exploded in recent years, foregrounding the need to unsettle Western conceptual and social orders. Decolonization is high fashion.[2] Often, however, everyday enactments and infrastructuring of this work reproduce figurations of the academic as a sovereign, proper-tied subject, deserving—through Lockean hard work—of exclusive rights. Academics speak against academic coloniality while further entrenching colonial structuration through their own affective and organizational investments—not always of their own volition. Meanwhile, ripple effects around the world continue with little comment.

This is yet another legacy effect, another way postcoloniality operates, and our concern here is with ways academia continues colonial structuration even when the dominating structures of academia are at the center of concern. The legacy is sustained through the colonial logics and infrastructures that underpin how scholars think about, produce, and (decline to) share their research—often encouraging "extroversion" (Connell n.d.; Hountondji 1990, 1997, 2009), tangling academics in commercially powered loops. Importantly, solutions are difficult if not impossible to locate through conventional frames—that reify opposition between centers and peripheries, for example, between academic and nonacademic spaces, and between academic rigor and play outside disciplinary lines. Colonial academic logics are often embodied by individual scholars but are important to understand structurally as built into academic cultures, institutions, and infrastructures. These logics also extend across borders, often operating with special force within universities in the Global South. Constructs of research competency, productivity, significance, and impact are homogenizing—and accelerating—despite vast differences in context both within and beyond universities, for example. Yet societal needs for research are of course different; the lives of academics in different places are also far from equal. From salaries to space, to access to books and journals as both readers and authors, to ways research—often barely—fits alongside teaching and service, academic conditions of production vary enormously. Paradoxically, academics in emerging economies often have less research time, space, and support than their northern counterparts, despite fanfare about the need for innovation and knowledge-based economies. Explicating these differences through lived experiences is a way to understand them in both particular and general terms, and as matters of money and time, technology and control—practicalities that are often unaddressed in decolonizing fabulations.[3] These practicalities tether academia to coloniality materially and psychically, calling for sustained analysis.

The chapter speaks to the central argument of this book in two ways. In the first part of the chapter, we describe what geographically decentralized academia looks like in practice—without adequate infrastructure to support the decentering of entrenched constructs of what is valuable in academia. The first segment describes the everyday life of internationalization in Indian universities. The second segment describes the contradictions of international academic publishing in Ecuador. The third segment describes ways contract time supports—and creatively constrains—research in Kenya. In the second part of the chapter, in contrast, we offer examples of academic decentralization that are, by design, counterhegemonic. In Leandro Rodriguez Medina and Sandra Harding's terms (introduction, this volume), they are both decentralized and decentered. Following Kaşdoğan's earlier research on scientists in provincial universities in Turkey, our focus is on ways academia on the so-called periphery (in Ecuador, Turkey, and Kenya) have built spaces for creative academic knowledge production, against many odds.

Together, the two parts of the chapter build toward an argument that we need to remoor academia, persistently loosening ties to colonial logics and infrastructures, slowly weaving alternatives.[4] Such weaving has many threads, organizational and institutional, technical and political, economic, rhetorical, and psychic, as well as conceptual. We work with the idea of unmooring to supplement thinking about decentering and decentralizing academia, foregrounding practical, organizational, and infrastructural aspects of the work. This, in turn, points again to the need to reconfigure ways academic work is recognized and evaluated. Habitual privileging of the conceptual over the organizational—encoded in splits between scholarship and service in academic accounting, for example—will need to be displaced. Talk about decolonizing academia will need to be tied to practice and supported with new infrastructure.[5]

We write here in a collective voice, in part as an experiment in writing that decenters single authors. We also, however, hope the diversity of our experiences from different communities of practice comes through. Aalok Khandekar, based in Hyderabad, India, describes the promises and paradoxes of transnational collaborative research relations, focusing on the dynamics of internationalization in Indian academic institutions. Angela Okune is a cofounder of the Research Data Share Working Group, formed to consider the kinds of academic practices and infrastructure called for amid Kenya's progress as the "Silicon Savannah." Maka Suarez, a cofounder of Kaleidos at the University of Cuenca in southern Ecuador, reflects on her experience building experimental spaces within the many constraints of Ecuadorian academia (Arias-Buitron et al. 2021). Duygu Kaşdoğan writes from her experience doing research alongside scientists in Turkey, then helping build IstanbuLab as a space beyond the university for academic work (Erol, Kaşdoğan, and Narin 2018). Kim Fortun writes across many years of work observing the need to build new transnational knowledge infrastructures and capacities, focusing especially on the challenges of environmental health and governance (Fortun 2021; Fortun et al. 2021).

Entanglements

In recent decades, a growing concert of voices has called for the decolonization of academic research both within and between countries.[6] This has created new opportunities for geopolitically marginalized scholars through international funding, mandated roles as collaborators, and invitations to publish in highly ranked journals (some open access), ostensibly opening a window for epistemic decentering. Many of these opportunities, however, draw researchers into a heavily commercialized academic system laced with colonial logics. It is a system rife with double binds for all scholars, undermining the promise of the university as a site of reflexive knowledge production and circulation, often prey to preset research agendas not always aligned with local needs. In what follows, we share stories of how this unfolds in India, Ecuador, and Kenya, drawing out multiple, intersecting aspects of colonized academia. By showcasing these challenges, we also open space for considering, in the next section, alternative possibilities.

Internationalizing Contradictions: Metrics and Academic Worlding

International visibility has become an increasingly important goal for many universities around the world, training eyes on international rankings and the metrics they are based on.[7] Faculty publications and evidence of international collaboration are key, as is the international diversity of both students and faculty. In the last decade especially, for example, there has been pressure to attract international students to Indian universities, supported by scholarships granted by individual institutions as well as the national government through a variety of bi- and multinational funding initiatives.[8]

Academics, too, welcome the opportunity to broaden and internationalize their work through collaborations. These opportunities for internationalization, however, often reentrench what we have referred to here as colonial academic logics (even while opening up new opportunities). How well Indian universities fare in various international university ranking schemes, for instance, is discussed in national newspapers every time a new set of rankings is released. Institutionally, there is increasingly a strong emphasis on publishing in journals credentialed by indexes such as Scopus and the Web of Science (Hazelkorn and Mihut 2021). Such commercially driven metrics act as organizing logics of academic production and become entrenched in everyday interactions, where journal impact factors and h-indexes come to be matters of routine discussion and markers of scholarly excellence that influence hiring and funding decisions.[9] And while many faculty at Indian universities experience these new demands on them as oppressive and push back against the quantification of academic productivity, the corporate consolidation (between publishing houses, indexing services, and ranking systems) of the metrics through which academic value is being articulated and measured remains largely unrecognized (Muthamilarasan and Prasad 2014).

Such moves toward internationalization are rife with contradictions. On the one hand, they open up possibilities, especially for scholars in the humanities and social sciences, that are difficult to imagine otherwise, given the privileging of science, technology, engineering, and mathematics (STEM) research in India's national funding system. Consider, for example, Khandekar's ongoing participation in an international research project focused on climate change adaptation and mitigation strategies in Hyderabad, India. Generous funding, made even more lucrative because of favorable currency conversion rates, has allowed him to establish a large and interdisciplinary research team. Many on the team aspire to use their time on the project to gain hands-on training with various tools and techniques—including, for example, designing and conducting surveys, ethnographic analysis, generating and analyzing spatial data, and mapping and modeling neighborhoods—and experience in academic publication that will hopefully allow them to secure positions in competitive doctoral programs internationally. The funding also allows his team to conduct workshops and create internship opportunities for students at local educational institutions as well as to imagine public-facing outputs that can help generate and amplify urgently needed discussions about climate change in the region in a subcontinent that has been repeatedly identified as a climate change hotspot. International projects, thus, open up possibilities for creating capacities locally.

An important dimension of Khandekar's participation in the climate change project is comparative: the goal is to understand climate impacts and responses across cities in the Global South. Given shared dynamics in underlying processes of urbanization (the dominance of informality, for example) in relatively resource-poor settings, knowledge sharing across different southern city-based teams promises significant possibilities for mutual learning. Regular exchanges between researchers on the team, both northern and southern, further supports collaborative learning, opens up avenues for participation in a diverse range of initiatives across disciplinary boundaries, and also establishes a network to potentially leverage for future funding (and other) opportunities.

Reliance on these funding mechanisms, however, produces its own peculiar dynamics, subjecting Indian institutions to agendas and audit cultures that are not necessarily aligned with domestic needs and priorities. In Khandekar's current project, for instance, it took more than a year (in a three-year project) for all international partners to agree and become party to the same funding contract. Funds could not be mobilized prior to this, making it impossible to hire project staff during that period. The project's start date coincided with the first wave of COVID-19, necessitating additional strategizing on behalf of the project team on how best to realize the project's commitments. Among other things, the contract stipulates that the legal jurisdiction for enforcing the contract is located in the funder's legal system, a particularly thorny matter for Khandekar's home institution, which is a centrally funded autonomous institute of the Government of India. The funding contract also stipulates that project expenses, by default, are in reimbursement mode, demanding extensive documentation of all expenditures incurred. To implement the process otherwise, that is, to receive payments in advance of spending, requires additional bureaucratic maneuvering, subjecting Khandekar's institution to additional due-diligence scrutiny, operationalized in the first place through a demand for a large number of different documents that do not always necessarily exist in particular formats (which vary depending on the funder anyway) or using requisite terminologies (that are very context specific). Suffice it to say that in this instance, the funder's additional demands that Khandekar's institution would need to satisfy in order to receive advance payments seemed daunting enough that the institute has foregone this possibility and is paying for the project and getting reimbursed for it once every quarter, but not without first making Khandekar personally liable in case of nonpayment. Thus, the very same favorable currency conversions that make the project lucrative also place considerable financial burden on the project partners—asking that they first spend large amounts of money and then work through bureaucratic and audit processes of the funders in order to receive the funds contractually allocated to them.

Given the emphasis on greater internationalization, Khandekar's experience is not unique by far. For every project that is funded by a different international funder, contracts, bureaucracies, and audit mechanisms need to be figured out anew, while still remaining compliant with domestic audit requirements as well. Like many other universities globally, too, administrative staff responsible for processing these projects are few. Inevitably, then, a large chunk of the administrative work for projects is further pushed down onto project teams themselves.

Even so, access to such opportunities is regulated through already existing academic structures: only some institutions recognized nationally as being reputable (as evidenced through university rankings) are deemed deserving to participate in internationalization initiatives, reentrenching hierarchies between academic institutions domestically.[10] Measuring up to valuation systems based largely on research productivity metrics is also key in order to successfully secure international funding, making publication in highly ranked journals a key priority for researchers. While this is not a problem by itself, such internationalization further entrenches a particular model of research and higher education as the ideal. Critiques of research evaluation along these lines are now well established. It is worth emphasizing in the context of the Indian university that it has been an important space for articulating demands of social justice in a deeply unequal society. Such alternative articulations of value are, in effect, rendered invisible in discussions over internationalization. Thus, even as academic internationalization processes are rife with rhetorics of and possibilities for collaboration and decolonization, the institutional mechanisms through which they are actualized also reinforce imperial relations of knowledge production (see also Çetinkaya 2017).

Consider what publishing and scholarly success look like from places like Ecuador (though what we describe happens in many other places).[11] Maka Suarez served as principal investigator (PI) for a team of five Ecuadorian researchers; the empirical collection had concluded, and an article was written, reviewed, and accepted for publication in a highly ranked Elsevier journal. This was good for the authors and for their university, which had hired them because of the potential for international publication—setting them apart from other faculty who had much heavier teaching loads. It nonetheless surprised the authors when they were asked to pay USD 3,380 to make their article "gold open access," removing all paywalls that would otherwise limit access to it (fig. 6.1). Even more surprising, however, was that both their home (public) university and a collaborating, private university in Ecuador agreed to pay this fee, without question.

Public universities in Ecuador are fully funded by the state and are free for undergraduate studies. Private universities, on the other hand, are a costly investment that few families can afford—even through indebtedness. For both these institutions, however, Elsevier's publication fee represents a hefty investment that is, needless to say, redirected from other demands. In the case of the public university, where malfunctioning electronic equipment, broken light bulbs, lack of access to books, and failing Wi-Fi services are the norm, it is an understatement to say that a payment of over three thousand dollars for an open access article is egregious. For a private university, the fee represents half the yearly cost of attendance for one student.

Invoice Address

Ecuador

Invoiced by: Elsevier B.V.

Total payment due	
	Price (excluding taxes)
	USD 3,380.00
Tax	USD 0.00
ⓘ Tax amounts are indicative and will be confirmed on the invoice	
	To pay
○人	**USD 3,380.00**
	Total price

Figure 6.1 Invoice from the publisher Elsevier to an Ecuadorian university.

Despite the costs of open access publication in this case, it could be said that it made sense since neither of the universities involved subscribed to the journal in which Ecuadorian research was being published. That, too, would have been too costly (Knöchelmann 2019). But this solution only feeds the problem, and a key academic publishing paradox.

In the name of leveling the playing field, there is an important, growing commitment to both open access scholarly publishing and research data sharing (Nelson 2022). The infrastructures supporting this, however, often reproduce entrenched academic privilege (Eve and Gray 2020; Moore 2021). The author-pay model for open access publishing—as illustrated in the Ecuadorian example—is one, particularly obvious example.

Open access in its current formulation does not fundamentally change underinvestment in academic institutions in the Global South, many of which have been gutted by structural adjustment policies. Through open access initiatives, scholarship becomes accessible, but what counts as open and who is able to publish openly is still delimited to those inside elite enclaves (particular countries, like Germany, for example, or institutional systems, like the University of California; Brainward 2021; Knöchelmann 2019). This does not promote a decentralized, bibliodiverse publishing system. Elites still have privileged access to both publication venues and published research products—and this is accelerating: Taylor and Francis, for example, recently announced that authors could be published more quickly—by paying more.[12] It thus is not surprising that recent developments toward open access have alienated many academics and publishers in the Global South, who have argued that both open access and open data in their current formulations could further marginalize them (Okune et al. 2021; Okune 2020; Knöchelmann 2021).

At the same time, the sheer labor of open access and more equitable forms of academic publishing is poorly supported and largely unaccounted for in academic productivity reviews. Open access remains largely underfunded by universities and academic societies, even while needing to live alongside (if not compete with) much better-funded corporate publishing. Further, many of these publishing corporations are located in the Global North, publish mostly in the English language, and are fully disconnected from local universities (not to mention local struggles). Many also have a reputation for unfair labor practices—beyond the fact that they don't pay for key product inputs (the academic articles themselves, and their reviews). Profit margins are, unsurprisingly, very high, sometimes near 40 percent (RELX 2019).

There are also other contradictions built into mainstream, largely corporate academic publishing. Journal impact factors, bibliometric data, and, in turn, university rankings are also generated by many of these same commercial corporations, for example (Posada and Chen 2017; Chen and Chan 2021). The publishing industry has also consolidated.[13] According to a study published in 2015 (based on a data set ending in 2013), the social sciences had the highest level of concentration, with 70 percent of papers published by the top five publishers. By 2006, five companies (Reed-Elsevier, Wiley-Blackwell, Springer, Taylor and Francis, and Sage) accounted for more than 50 percent of overall published academic output, up from 20 percent in 1970 (Larivière, Haustein, and Mongeon 2015).

Academic publishing in established, elite venues thus cannot be thought of as a means to decenter and decentralize knowledge and academia. This can only be achieved through the harder work of shifting research practices and cultures, and building supporting infrastructure, enacting rather than just claiming decolonized approaches.

Contract Time: Academic Coloniality beyond
the University

Logics of coloniality in academia are not restricted to the ivory tower. In Kenya, for example, the majority of research work takes place outside of university structures. Low-paying lectureships in universities often prompt academics to take on external consultancy projects for a diverse array of organizations such as the United Nations, World Bank, or other for-profit or nonprofit organizations.

"It's like these people don't realize that we budget by the hour!," said an exasperated researcher at iHub to Angela Okune after a particularly long Skype call with a collaborator. iHub is a flagship Kenyan technology co-working space that became a prominent fixture in Nairobi after 2010, in step with Nairobi's emergence as a tech-for-development hub often referred to as "Silicon Savannah" (Okune and Mutuku 2023). "[They don't understand that] we are not full-time academics that can just work on one project as long as we want without regard for how many hours we put in. I don't think they know that we juggle ten projects at one time!"

Work at iHub and in many other settings where contract academic work is the norm illustrates the day-to-day ways that structures of dependence are reproduced in practice. The time-keeping aspect of contract work is one part of this. So, too, is the pull to be helpful locally while also informing policy and being academically rigorous. Multiple, often competing evaluative criteria are in play. The case of iHub is again illustrative. Many projects were done in collaboration with local and international academics, since half of iHub's funding came from philanthropic donors or development aid organizations and the remaining half from market research work for private corporations. iHub nonetheless adopted a business model more like a private consulting company than an academic department. That is to say, researchers had to count their hours, allocating just the right amount of time to different projects, always mindful of the amount of funding available to cover the time.

While working at the research department at iHub Kenya from 2010 to 2015, Okune heard numerous complaints about team members feeling burned out. In hindsight, she realized that there were just not enough hours in the day to do all of the work that iHub had committed to do. But iHub couldn't hire more people because then it would have to find more projects to keep paying them. iHub needed to keep projects going so that existing staff could be paid, but because margins on projects were too small to cover all overhead costs, they were stuck in a loop. It was a nefarious cycle—experienced in many organizations in Nairobi, and far beyond.

At Akamai Lab in Nairobi, another research organization where Okune (2021) did research, the time management was even more granular and intense—micromanaging staff time down to the hour (a technique iHub had moved beyond).[14] One researcher showed Okune what this looked like in color on a Google calendar, with differently colored blocks covering thirty-minute periods. "So I can remember what project I was working on when I have to fill in my time sheet," she explained, grimacing slightly. She then logged into the system and showed Okune the combination of multiple HR and time-tracking softwares that she filled out regularly. Each hour was supposed to be counted toward a billable project. "Which project should we put this meeting under?" was a question that was often raised on the company Slack before or after a company-wide general meeting.

This contract time is a direct lack of overhead support and leaves no time for the nonproject work needed to thoughtfully build knowledge infrastructure.[15] The practical is delimited by a restrictive funding model. This experience from the Kenyan not-for-profit research scene helps shed light on how the funding of research determines the kinds of activities that are supported and the kinds of activities that are underinvested in. It also reveals why genuine collaboration with those working outside of the university remains structurally difficult. Decentering and decentralizing knowledges requires grappling with the ways in which academic professionalism relates to colonial time and a work style that centers and values certain kinds of productivity.

Spaces Between

"For me, everywhere could be a lab; after all, the lab is in my mind," said a biochemist in response to questions about the hardship of not having a lab of his own in a provincial university in Turkey. Duygu Kaşdoğan heard similar statements from other scientists throughout her fieldwork in Turkey's provincial universities in 2015. Repeatedly, they told her that lacking scientific infrastructure didn't keep them from doing science—that they consistently found a way. They reused equipment found in storage spaces; they found ways to keep up with global trends, even without access to most academic journals; they learned to work in step with the claim that "here in Turkey, nothing endures." Kaşdoğan both documented and was inspired by this work.

Stories of staying with science and academia against extraordinary odds abound in other places, too. Noémi Tousignant, for example, describes incredible persistence and commitment among Senegalese toxicologists in the wake of structural adjustment, which gutted their labs and directed all their time into contract work. Despite the risks in tracking toxic exposure and their limited infrastructure, budgets, and staff, toxicologists moved between formal academic spaces, nongovernmental institutions, and private laboratories in order to build capacity for undertaking critical, Senegal-based research (Tousignant 2018). Sebastian Ureta (2021) describes similar persistence in Chile, where toxicologists practice what he calls *ruination science*, the ability or willingness to produce scientific knowledge despite ruination of scientific infrastructures.[16]

Our next section follows from these examples, describing various (sometimes completely unfunded) efforts to build alternative academic organizations and infrastructures, against many odds. The need for improvisation and adaptability cuts across our examples, as does the need for creative fencing with established bureaucracies and academic organizations. Alternatively, decentralized academic knowledge production also requires stubborn optimism and a deep sense of purpose.

IstanbuLab

Especially for academia, 2016 was a dramatic year in Turkey. The direct appointment of university rectors by the country's president destroyed "the last remnants of academic self-governance" (Erdem and Akın 2019, 147). Academic freedoms were further violated after a coup attempt and subsequent declaration of a state of emergency. The concept of science was dropped from the name of the Ministry of Science, Technology, and Industry. Academics who signed petitions stating "We will not be a party to this crime!" (known as the "Peace Statement") were dismissed from their positions.

IstanbuLab emerged in this context, out of a need for breathing spaces outside the university, and as a response to the current environment "to undertake critical studies of science as well as supporting scientific inquiry" (Erol, Kaşdoğan, and Narin 2018). IstanbuLab is an independent volunteer-based research network for social studies of science and technology, bringing together academics, artists, and activists.[17] This was a shared commitment among the founding members of IstanbuLab, who at the same time had multiple other motivations as well.

Founding IstanbuLab member Duygu Kaşdoğan approached the work from her position at what can be considered a *taşra* (provincial) university, where teaching loads are especially high. Looking back after five years, Kaşdoğan recognizes that her motivation in helping establish IstanbuLab was at least twofold: she wanted to help preserve space for critical academic knowledge production, uncensored by authorities, attuned to local needs. She also wanted to create a space for herself and others that wasn't sustained by self-sacrificing scientific bodies. Historian Rebecca Herzig (2006) has written about the rise of an ethic of "self-sacrifice" in American science during the late nineteenth and early twentieth centuries. The self-sacrificing scientific body has a different genealogy in Turkey, where, at least since the founding of the Republic (1923), Turkish scientists have been lauded for overcoming political and economic barriers, sacrificing self for science and state. IstanbuLab, in Kaşdoğan's vision, needed to provide an alternative.

Whether counterhegemonic spaces outside universities inevitably reproduce self-sacrificing scientific bodies remains an open question. The pains and potential abuses of voluntary labor are well known. IstanbuLab constantly and creatively works against these. One strategy has been to foreground local and transnational collaborations. IstanbuLab has centered collective endeavor, opening spaces to unlearn and relearn what it means to produce and leverage knowledge—"activating arts of living and dying well in a damaged planet," as lab members often say (Haraway 2016; Tsing et al. 2017).

One collaborative project undertaken by IstanbuLab focused on the history of critical studies of the science-society nexus in Turkey, produced as an exhibit for the 2018 annual meeting of the Society for Social Studies of Science, an organization representing scholars and practitioners in science, technology, and society (STS) studies.[18] IstanbuLab's exhibit helped draw out the deep transnationality of STS, which hasn't been well represented in the field's leading journals (though this is rapidly changing, with the development of journals like *Tapuya: Latin American Science, Technology, and Society*, centered in Latin America; EASTS (*East Asian Science, Technology and Society*), centered in East Asia; and *Engaging Science and Technology Studies*, now led by a transnational editorial collective). By design, the 2018 STS Across Borders project put IstanbuLab's exhibit alongside and in conversation with STS communities and scholarship in other regions. Importantly, many of the exhibits (focused on Turkey and Africa, most notably) developed their regional histories of STS despite the relative newness of STS as an organizing concept in these regions. These exhibits created bodies of work that previously were less recognizable as such. The STS Across Borders project—through the work of organizations like IstanbuLab—thus decentered STS (locating it beyond Euro-America) and decentralized it, helping build critical memory and recognitional and organizational capacity with and between STS communities literally around (rather than at the top) of the world.

IstanbuLab has again mobilized to build critical memory and recognitional and organizational capacities in Turkey in the wake of the catastrophic earthquake that hit the southeast corner of the country (and neighboring regions of Syria) in February 2023. Following the earthquake, the (still very authoritarian) Turkish government tried to fully control both the recovery process and narratives about the recovery—going so far as to repress mutual aid activities. Journalists and diverse scientific and professional societies were also silenced. The importance of preserving a people's history of the earthquake, curated to scaffold ongoing, desperately needed public discussions about the 2023 earthquake, as well as past and future earthquakes, in Turkey quickly became clear. IstanbuLab has invested deeply in this work, again both documenting and building critical knowledge capacity in Turkey by mobilizing both transnational and local collaborations (making use of open source digital research infrastructure built by the Disaster-STS Network, for example). The work is both decentering and decentralizing.

Kaleidos

In Ecuador, entry into academia is difficult for many reasons—despite a government program designed to support Ecuadorian students abroad, then bring them home to spark Ecuador's next stage of development (ICEF Monitor 2014). Disciplinary identities and divisions are strict. Permanent positions in public universities function within clientelist networks that intersect with kinship, class, race, and gender. Permanent positions in private universities are nearly nonexistent. There are few tenure-track positions in Ecuador's higher education, and those at public universities are often earmarked before they hit the job market. The highest authorities in public universities are elected from the votes of professors holding permanent positions (along with permanent staff and students, though their votes weigh much less). This effectively results in each new permanent hire becoming also a new vote. So even though, in theory, permanent positions are filled through public competitive examinations, in practice, they work within complex networks of patronage meant to increase institutional power. As one senior academic explained, "Yes, they [permanent positions] are supposed to be open to competition but there is much subjective choice. Interviews, which are highly marked [given greater weight], are not public so it's very hard to dispute the [overall] results."

Kaleidos Center for Interdisciplinary Ethnography was created within and as a response to this context, and Maka Suarez is one of its cofounders. Launched in April 2018 at the University of Cuenca in southern Ecuador, with initial support from FLACSO-Ecuador (Latin American Faculty of Social Sciences–Ecuador), Kaleidos is an experimental academic space built to break divides between both disciplines and generations, encouraging transnational as well as local flows of ideas. Organizationally, it sits between different universities, civil society organizations, government institutions, and private companies. Cofounded by three Ecuadorian academics, it moved between three different universities in its first five years of existence. Kaleidos has faced many challenges, including what Hebe Vessuri terms "contested conviviality." Contested conviviality, Vessuri explains, refers to the need for "negotiating spaces in different contexts, which are often precarious, asymmetrical and sometimes risky" (2019, 27).

Creating a new research center in this academic context was based on negotiations at varying management levels. At every university with which Kaleidos was affiliated, the highest authorities were always involved in opening up the space for Kaleidos to exist. The center could not have existed otherwise but, at the same time, it created unsettled feelings among colleagues who had long been part of the university, particularly among parties opposed to current authorities. Younger academics sometimes felt threatened by Kaleidos; more senior academics sometimes disagreed with the epistemological vision. Old disciplinary understandings left little room for innovation.

Kaleidos brought new understandings of anthropology and STS to Ecuador, which was met with much enthusiasm from younger generations but also with suspicion, wariness, and mistrust from colleagues with different academic agendas. Kaleidos eventually found its footing and a welcoming home in the University of Cuenca's Engineering Department. Colleagues in the hard sciences had little understanding of anthropological or STS methodologies or theoretical discussions, but the certainty they held in numbers (and the positivist method more generally) meant that Kaleidos's interdisciplinary ethnographic proposals were met with curiosity rather than suspicion, even if there was initial reservation. Rather than being convinced by our methods or theories, engineers were interested in collaborations. They saw potential in Kaleidos's research process: extramural funding, international academic partnerships, high-impact articles in our respective disciplines (this meant mostly English-language journals). To Kaleidos, cooperation was attractive because it opened up new areas of study (air pollution, solar energy, robotics, and artificial intelligence), a pool of students eager to learn new skills (particularly those who wanted more critical thinking), and peace of mind to conduct their own projects. Ultimately, Kaleidos was assigned a large office space far from the main campus, which was refurbished with recycled furniture, equipment, and eclectic paraphernalia.

Kaleidos was a success, at least for a few years, even if it was unstable and vulnerable. Its loose link to the Engineering Department provided some perks—little bureaucratic responsibility, no department meetings, next to no interference with its work. It also meant Kaleidos researchers were vulnerable to external shocks: they had no guaranteed teaching time, forcing them to jump from one department to another each semester in order to complete assigned teaching loads, always in distress up to the last minute to complete their schedules. Much of this was dependent on personal relations rather than institutional processes, and Kaleidos researchers were on temporary contracts—which meant a double vulnerability.

In 2020, during the COVID-19 pandemic, there were large cuts to educational budgets in public institutions, including the University of Cuenca.[19] In a mix of funding shortage and long-standing patron-client relations, authorities chose to fire many of those on temporary contracts, including Kaleidos cofounders. As an epistemic project, Kaleidos disrupted dominant understandings within the field of social sciences; here our ambiguous location within the hard sciences provided little shield in the face of contract negotiations. Kaleidos struggled to stay afloat, moving to a new private university in a different city, which felt like starting all over again. It was hard and exhausting, and it happened in the middle of the pandemic. This switch, though fortunate as it meant new jobs, took a heavy toll on the team and the infrastructure they had managed to carefully piece together in the previous years. The center did not end but was reduced to its minimal expression. The new university had very different requirements, however, and Kaleidos found itself in uncharted territory, one in which anthropology—as a discipline—was not needed or desired. Rather, a clear map was set up for Kaleidos, rather than with Kaleidos—one that included supposedly universal academic metrics. The arrangement did not last. After six months everyone agreed that the private university was not the best location for Kaleidos.

Starting in September 2021, Kaleidos returned to the University of Cuenca. New authorities have reaffirmed their interest in the center's research project in the short and long term. Many vulnerabilities remained, but a new chapter in the short history of Kaleidos opened. Kaleidos researchers still believe that research spaces like Kaleidos have great potential for breaking old barriers within universities, particularly for young academics and those trained in different academies, and for reimagining research in increasingly budget-constrained realities.

Research Data Share

Because it was founded by and for white settler colonialists in the 1950s, discussions about decolonizing the African university in the 1960s grappled directly with its colonial legacy, questioning how to reorganize the university to center African knowledges and more closely reflect the experiences of and for African students. In 1972, for example, the trio of Ngũgĩ wa Thiong'o, Henry Owuor Anyumba, and Taban lo Liyong famously proposed abolishing the English Department at the University of Nairobi to make space for literary forms and aesthetics rooted in Kenya rather than outside (Musila 2019; Gikandi and Mwangi 2007). This was part of broader attempts to reorient "extroverted" African knowledge production (Hountondji 1990). The time and space allowed for these important discussions of restructuring was, however, short-lived. In the 1980s, structural adjustment programs imposed by Bretton Woods institutions resulted in the Kenyan government reducing per capita expenditure on various social services, including education. Previously, state-subsidized university education came under a cost-sharing plan that made the cost of education increasingly unaffordable for students from poor backgrounds. Kenyan scholars have detailed some of the cascading effects of the introduction of these education cost-sharing plans, especially the increased need for university students to have income-generating jobs in the midst of their studies (Kamau 2005; Muyia 1996).

Today, many Kenyan students self-fund their education and research. In addition to struggling to pay their way through postgraduate study programs, not to mention using personal funds to cover research costs, these students also lead fragmented, stressed lives trying to raise funds to cover their everyday expenses and personal obligations. This experience of hustling within the university appears to be true of both students and lecturers. Overburdened lecturers supervise a large number of students, so students are unable to receive the mentorship they seek. Individual lecturers themselves are often campus-hopping from one temporary position to the next, or working on multiple consulting projects in order to generate enough income to survive. These lecturers similarly struggle with finding the time and support to conduct their own research. Contract time doggedly follows researchers as they move between, within, and outside the university walls.

The Research Data Share (RDS) platform was designed to counter this context. Established in 2018 to support Angela Okune's dissertation research, it has grown into an experimental space for considering what kind of research, intellectual life, and knowledge infrastructure are needed in Kenya and Africa, both expansively conceived. It is an intervention, not an end in and of itself—a way to make space to reflect on the structures under which knowledge is being produced in Kenya. The digital platform was set up as an instance of the open source software called the Platform for Experimental Collaborative Ethnography (PECE). The web-based data platform houses primary data from Okune's doctoral research as well as found artifacts on research and a COVID-19 project that emerged with the pandemic. The platform is incomplete, which is to say, it does not house cleanly completed outputs or finished publications but instead has been added to in bursts and spurts, as creativity hits and time allows. Recognizing the limitations of researchers working under contract time in other spaces, RDS has established social norms and infrastructure that support group members to come and go, recognizing the challenges of balancing voluntary labor with other commitments and challenges.

The platform has served as a focal point for group discussion, fostering and creating space to figure out what the space of and for critical knowledge production in/about/for Kenya needs to be. The questions considered are many and diverse: What kinds of research data are important to archive for Kenyan futures, for example? How should such data be collected, stewarded, and maintained?

Members of RDS are located within and outside of university structures; they are tied together through a shared interest in research collaboration. The work is slow, complex, and messy, but the group has found inspiration in the new connections and learnings they have crafted together. For us, RDS exemplifies the technical and social infrastructures, processes, and labors through which decentralization can be pursued.

Remooring

In writing about how communities rebuild after disaster, sociologist Tyson Vaughan looks to the example of Minamata, Japan, a place devastated by corporate mercury pollution that moved invisibly and insidiously through food chains and bodies, bioaccumulating, ruining functional capacities.[20] Extensive brain damage was both grossly literal and a metaphor for broken technical and political, conceptual and cultural systems.

Vaughan's focus is on what happened in the aftermath, foregrounding the way people in disaster settings envision and work toward recovery. In Minamata, he explains, people "use the term 'moyai-naoshi' (remooring) to capture their approach to recovery" (Vaughan 2021, 197). To remoor is not to build back to the way things were, but to configure things differently, to reorder relationships. In the local dialect (not standard Japanese), *moyai* literally means to tie boats together (rather than to a dock or anchor).

Naoshi refers to repair, or redoing something to get it right. Collaboration is thus key, as is reflexive recognition that old structures that produced disaster shouldn't be relied on going forward.

Recognizing that the work of rebuilding academia is very different from the work of rebuilding in the wake of environmental disaster, we nonetheless find the idea of remooring compelling and useful for advancing the argument we present here.[21] Remooring, as described by Vaughan, involves new ways of organizing, leveraging lateral rather than hierarchical relations. It is a way to think about what epistemic decentering and decentralization look like in practice.

As Rodriguez Medina and Harding convey (introduction, this volume), epistemic decentering—unsettling entrenched epistemes through pluralization—requires institutional support and change, what they call epistemic decentralization. Remooring aptly describes what it looks like to enact such institutional support and change, recognizing the importance of changing many things at once—conceptual, political, economic, technical, and so on, establishing many new ties.

Vaughan goes on to explain that remooring is set against entrenched, statist approaches to disaster recovery, which reliably and ironically work to reestablish the predisaster status quo, with success measured through predisaster metrics (even though the views of the world afforded by these metrics may well have contributed to the disaster). Vaughan (2021) also foregrounds postdisaster doubling down of predisaster economic and technological commitments—to neoliberalism as an economic plan and to surveillance as supporting knowledge form, for example.[22] Remooring has to be strategized within such a doubling down. There are lessons for us here, too: Academic remooring also always works within and against entrenched hierarchies and is inevitably judged with entrenched metrics. In many contexts, retrenchment is a paradoxical effect of reaching for something new (as with metric-driven internationalization in Indian academia, for example).

IstanbuLab, Kaleidos, and Research Data Share all remoor, attuned to their very different contexts but also strengthened by their connections. They operate independently and are stronger because tied together. On their own, and collectively, their everyday enactments of academia shift entrenched structures and unsettle usual ways of thinking about the nature of academic work, where it happens, and how it should be evaluated.

In building these alternative academic spaces, many questions stay with us: How can we rebuild educational and research capacities wrecked by structural adjustment without determining their form and direction in advance (as a condition of their funding, among other things)? What kinds of transnational funding and collaboration not only tolerate but encourage peripheral vision and explanatory pluralism? How can archives and other research resources be built so that both their content and form, by design, constantly unsettle established category schemes? What forms of academic life can skirt deeply entrenched proprietary habits within academia itself? How can academic work address the array of vectors and dynamics that reproduce colonial hierarchies and exclusion?[23]

Like the mercury that poisoned Minimata, coloniality has operated insidiously in academia, creeping into the imaginations, comportments, and practices of academics themselves, settling in technical infrastructures. Our challenge, then, as in Minamata, is to rebuild differently, changing organizational as well as conceptual habits, enacting academia anew. Epistemic decentering and decentralization have to be infrastructured.

Notes

1   Black feminist theorist Katherine McKittrick articulates a sobering analogy, pointing to ways thinking and acting in terms of owning, possessing, and excluding in academia reproduce both colonial and plantocentric (plantation) logics. McKittrick describes a sharp moment of realization she had many years ago at a dissertation workshop, when students were talking about how "some senior scholars claimed they 'owned' certain slave archives." McKittrick goes on to recall

> feeling really nauseous and confused about how this research was being described through the practice of ownership—the re-proprietorship, actually—of Black enslaved stories in this way. I cannot get this out of my head. I guess the senior scholars were saying, "That's my slave, I found her first," with a real enthusiasm. I was very rattled. This was a cautionary tale for me! . . . To own Black thought and Black methodology closes down curiosity. It belies what Black methodologies are tasked to do: which is to work, read, think, create across different sites and knowledges and histories, in order to undo plantocratic and colonial logics that thrive on owning, possessing, having, excluding, extracting. (McKittrick 2021)

See also Sabelo J. Ndlovu-Gatsheni's (2021) blog post, "The Cognitive Empire and Gladiator Scholarship."

2   In a 2022 article, Perry Guevara and Amy Wong write, "In the last decade, 'reckoning' literature in higher education has become a genre on its own. In the wake of groundbreaking studies, such as Craig Steven Wilder's *Ebony and Ivory* (2013), American colleges and universities have had to

acknowledge, for the first time, their culpability in institutionalizing and profiting from slavery."

3   Our argument is akin to (and also different from) Eve Tuck's and K. Wayne Yang's (2012) argument that "decolonization is not a metaphor." Tuck and Yang write against the adequacy of "decolonizing the mind," emphasizing that decolonization has to be understood in material terms—as giving back Native land. Their argument is important and complex and raises many questions. See, for example, a critical reading of Tuck and Yang, emphasizing that "to be anti-metaphor is to be anti-Black" (Garba and Sorentino 2020, 776). Here, Tuck and Yang can be invoked to emphasize the need for organizational work that refigures subject and social formations. Decolonizing, as they say, will take more than words (though words—and metaphors—are also critical, in our view). Our emphasis is on the organizational work that must be done to enable different ways of languaging and enacting the world.

Note also Ngũgĩ wa Thiong'o's (2011 [1986]) impressive *Decolonizing the Mind*, and more recent commentary by his son, Mukoma wa Ngũgĩ (2018). Asking what *Decolonizing the Mind* brought to the table, Mukoma wa Ngũgĩ explains that "it tied language and culture to the material work of both colonization and decolonization. . . . It also examined the close relationship between language and culture." Mukoma wa Ngũgĩ (2018) goes on to explain that there is much (organizational) work ahead in the reawakening of African languages, providing these details:

> In Kiswahili, which has an estimated 100 million speakers, there are only a handful of literary journals. And prizes for Kiswahili literature are not more than five. For Gikuyu, my mother tongue spoken by close to 7 million people, I can name only one journal: Mutiiri, launched by my father in 2000 as a print journal, and now found online. There are no literary prizes associated with the language. Publishers of literary texts in African languages outside of South Africa are few and far between. I do not know of a single journal that produces literary criticism in an African language. Or any residencies that encourage writing in African languages. The point is, for a population that will soon reach 1 billion people, spread over 55 countries, even 100 journals and literary prizes would still be pitifully inadequate.

4   Sørenson and Traweek (2021) describe this kind of weaving and organizational work as the "meshwork" of academia, carried out both alongside and in opposition to the work of building and maintaining hierarchies in academia.

5   There is a very long and rich history of effort to preserve the university. See, for example, John Henry Newman's (1982 [1852]) *The Idea of a University*, where he argues that higher education shouldn't be oriented or controlled by authorities or the ends they imagine. We also build on

rich recent work in critical university studies (Sørenson and Traweek 2021; Scott 2019; Bhambra, Gebriel, and Nişancıoğlu 2018; Biesta 2017), academic imperialism (Alatas 2000, 2003), epistemic injustice (Bhargava 2013; Fricker 2017; Ogone 2017; Pitts 2017), and decolonizing methods (Mignolo 2012; Ndlovu-Gatsheni 2015; de Sousa Santos 2018; Smith 2014; Soto Laveaga 2020; Rivera Cusicanqui 2012; Chen 2010).

6   See, for example, *Decolonizing the University* (Bhambra, Gebriel, and Nişancıoğlu 2018, 4): "We hope that a discussion of decolonising from the imperial centre—of which this volume is only one part—might help to re-veal something about the machinations of empire in general and the deeply understudied relationship between coloniality and pedagogy. In doing so, it also has the potential to open spaces for dialogue, alliances and solidarity with colonised and formerly colonised peoples, contributing to the making of 'a global infrastructure of anti-colonial connectivity.'"

7   Pressure to internationalize is happening in many places, and there is a growing literature that explicates its effects (Celis and Guzmán-Valenzuela 2021; Thondhlana et al. 2020; Rodriguez Medina 2018; *ICEF Monitor* 2015).

8   India's Ministry of Education, for example, administers the GIAN (Global Initiative for Academic Networks) and SPARC (Scheme for Promotion of Academic Research and Collaboration) programs. GIAN supports mobility of internationally renowned researchers into India by support-ing one- to two-week intensive courses cotaught by them and a collabora-tor in India. SPARC supports researcher mobility between collaborating institutions in and outside India. At last count, these programs support collaborations with institutions in about thirty countries, most of them based in Europe, North America, and East Asia (Brazil, Israel, Russia, and South Africa are exceptions).

9   An *h*-index is an increasingly commonly used metric for measuring re-searcher productivity and publication impact.

10 Academic hierarchies within a given national context are also powerfully reproductive. See, for example, Nicolas Kawa et al.'s (2019) analysis of continually reproduced inequalities between anthropology departments in the United States.

11 There is a rich body of scholarship that examines academia in Latin Amer-ica (Rodriguez Medina 2018), including threads focused on the region's complex publishing ecology (Beigel 2021) and on the development and contributions of the interdisciplinary field of science and technology stud-ies (Kreimer and Vessuri 2018; Kervran, Kleiche-Dray, and Quet 2018).

12 See Taylor and Francis's "Accelerated Publication" options (for a fee): https://taylorandfrancis.com/medical-publication-professionals /accelerated-publication/.

13 Reed-Elsevier, Wiley-Blackwell, Springer, and Taylor and Francis are the top four publishers across fields; Sage is the fifth for social sciences, while American Chemical Society is fifth for the natural and medical sciences.

14   Akamai Lab is a pseudonym.

15   Many research funding agencies do not allow operational overhead costs
      of a research project budget to run more than 13–15 percent of the overall
      research grant. So a USD 17,000 project, for example, might only contrib-
      ute a little more than USD 2,000 toward covering overhead costs. Overhead
      costs typically include core operational functions like office space,
      electricity, and water; administrative staff, including accountants and
      a finance manager; and technical infrastructure, including server costs,
      and software and hardware maintenance, among other things. There are
      few if any grants solely to cover operational and running overhead costs;
      each research grant must be tied to a project. And since each project only
      allows a small percentage of the grant to go toward overhead costs, you
      need many projects to be running simultaneously to cover organizational
      overhead costs. Not to mention the bureaucratic nightmare that getting
      paid via multiple projects can mean for researchers and staff.

16   Also see Pankaj Sekhsaria's (2019) *Instrumental Lives: An Intimate Bi-
      ography of an Indian Laboratory.* The literature on "southern Urbanism"
      also highlights the resourcefulness of people working in conditions of
      considerable deprivation. See, for example, Gautum Bhan's (2019) "Notes
      on a Southern Urban Practice."

17   See IstanbuLab, https://www.stsistanbul.org.

18   For an extensive discussion on the ways STS have been developing in
      Turkey as well as an analysis of IstanbuLab's activities deriving from this
      4S exhibition, see Alkan, Kaşdoğan, and Erol (2023).

19   In Ecuador, the pandemic legitimated long-planned austerity pack-
      ages that reduced public spending by USD 4 billion. Many of these cuts
      were already unfolding, provoking large waves of social unrest in 2019
      (Suarez and Núñez 2020). When Ecuador became the epicenter of the
      pandemic in March 2020, the government announced new economic
      measures that included a 10 percent cut to public education (nearly USD
      100 million; Torres 2020). The official narrative was that cuts were due
      to harsh economic circumstances due to the pandemic and a global fall
      in oil prices (Ecuador's main export). However, at the same time, the
      government announced its intention to prioritize debt repayments to
      the International Monetary Fund on a USD 4.2 billion loan contracted in
      2019 (Solano 2020). Taking advantage of the strict lockdown at the time
      and the inability of people to protest on the streets, an austerity package
      undermining public services was approved. There is little doubt that, even
      though the pandemic has greatly reduced economic possibilities, it has
      also become, at least in the context of Ecuador, a conduit for neoliberal
      projects that had previously met strong opposition on the streets. The
      COVID-19 pandemic provided the perfect political scenario where swift
      social changes were made with dubious or no real justification.

20   Vaughan (2021) has done extensive research on earthquake and tsunami
      disaster recovery in Kobe and Tōhoku, Japan. Vaughan went on to work

as a sociologist for the US Army Corps of Engineers Institute for Water Resources.

21   In a useful comparison of Winnicott and Lacan (noting the professional antagonism that their differences engendered), psychoanalyst Deborah Luepnitz points out that "the word 'analysis' comes from the Greek verb amakt im [analyein], meaning to loosen or untie. Lacan writes: 'Psychoanalysis alone recognizes this knot of Imaginary servitude that love must always undo again or sever' (1949, 7). For Winnicott, analysis may untie or free the True Self from its moorings in compliance" (2009, 974). Luepnitz's explication of the difference between Winnicott and Lacan, emphasizing the therapeutic value and significance of both, is also useful here—for thinking about how academic formations need to be both held (Winnicott) and interrupted (Lacan).

22   Vaughan's description of state tendencies in postdisaster contexts draws on David Edgington's (2010) *Reconstructing Kobe: The Geography of Crisis and Opportunity*.

23   There have been many efforts to theorize and enact problems of form— the way things are put together, the ways things are put into relation. Deleuze and Guattari (1987) provide a persuasive explication of the difference between arborescent rather than rhizomatic forms, for example.

References

Alatas, Syed Farid. 2003. "Academic Dependency and the Global Division of Labor in the Social Sciences." *Current Sociology* 51 (6): 599–613.

Alatas, Syed Hussein. 2000. "Intellectual Imperialism: Definition, Traits, and Problems." *Southeast Asian Journal of Social Science* 28 (1): 23–45.

Alkan, Aybike, Duygu Kaşdoğan, and Maral Erol. 2023. "Placing STS in and through Turkey." *Engaging Science, Technology, and Society* 9 (1): 104–24.

Arias-Buitron, Natalia, Grégory Deshoullière, Jorge Núñez, and Maka Suarez, eds. 2021. "Thinking with Decoloniality: Authorship and Collaborations in Neoliberal Times." Virtual issue, *JRAI*. https://rai.onlinelibrary .wiley.com/doi/toc/10.1111/(ISSN)1467-9655.decolonising.

Beigel, Fernanda. 2021. "A Multi-scale Perspective for Assessing Publishing Circuits in Non-hegemonic Countries." *Tapuya: Latin American Science, Technology and Society* 4 (1): 1845923. https://doi.org/10.1080/25729861 .2020.1845923.

Bhambra, Gurminder, Dalia Gebriel, and Kerem Nişancıoğlu, ed. 2018. *Decolonizing the University*. London: Pluto.

Bhan, Gautam. 2019. "Notes on a Southern Urban Practice." *Environment and Urbanization* 31 (2): 639–54. https://doi.org/10.1177/0956247818815792.

Bhargava, Rajeev. 2013. "Overcoming the Epistemic Injustice of Colonialism." *Global Policy* 4 (4): 413–17.

Biesta, G. 2017. "Don't Be Fooled by Ignorant Schoolmasters: On the Role of the Teacher in Emancipatory Education." *Policy Futures in Education* 15 (1): 52–73. https://doi.org/10.1177/1478210316681202.

Brainward, Jeffrey. 2021. "California Universities and Elsevier Make Up, Ink Big Open-Access Deal." *Science*, March 16. https://www.sciencemag.org/news/2021/03/california-universities-and-elsevier-make-ink-big-open-access-deal.

Celis, Sergio, and Carolina Guzmán-Valenzuela. 2021. "Internationalisation and the Global South." *Scholarship of Teaching and Learning in the South* 5 (1): 1–5. https://journals.uj.ac.za/SOTL/index.php/sotls/article/view/179.

Cetinkaya, Eda. 2017. "Turkish Academics' Encounters with the Index in Social Sciences." In *Universities in the Neoliberal Era: Academic Cultures and Critical Perspectives*, edited by Hakan Ergül and Simten Coşar, 61–92. London: Palgrave Macmillan.

Chen, George, and Leslie Chan. 2021. "University Rankings and Governance by Metrics and Algorithms." In *Research Handbook on University Rankings: Theory, Methodology, Influence and Impact*, edited by Ellen Hazelkorn and Georgiana Mihut, 425–43. London: Edward Elgar. https://zenodo.org/record/4730593#.YIwg3uspDOQ.

Chen, Kuan-Hsing. 2010. *Asia as Method: Toward Deimperialization*. Durham, NC: Duke University Press.

Connell, Raweyn. N.d. "The Cultural Cringe and Social Science." *Southern Perspectives* (blog). Accessed July 8, 2024. https://southernperspectives.net/tag/paulin-hountondji.

Deleuze, Gilles, and Felix Guattari, 1987. *A Thousand Plateaus: Capitalism and Schizophrenia*. Minneapolis: University of Minnesota Press.

de Sousa Santos, Boaventura. 2018. *The End of the Cognitive Empire: The Coming of Age of the Epistemologies of the South*. Durham, NC: Duke University Press.

Edgington, David. 2010. *Reconstructing Kobe: The Geography of Crisis and Opportunity*. Vancouver: University of British Columbia Press.

Erdem, Esra, and Kamuran Akın. 2019. "Emergent Repertoires of Resistance and Commoning in Higher Education: The Solidarity Academies Movement in Turkey." *South Atlantic Quarterly* 118 (1): 145–63.

Erol, Maral, Duygu Kaşdoğan, and Özgür Narin. 2018. "IstanbuLab: Building New and Excavating Old STS Infrastructure in Turkey." *Backchannels: Society for Social Studies of Science*, October 29. https://members.4sonline.org/news_archive_headlines.php?org_id=4S&sniid=34479631.

Eve, Martin Paul, and Jonathan Gray, eds. 2020. *Reassembling Scholarly Communications: Histories, Infrastructures, and Global Politics of Open Access*. Cambridge, MA: MIT Press.

Fortun, Kim. 2021. "Cultural Analysis in/of the Anthropocene." *Hamburg Journal of Cultural Anthropology* 13: 15–35.

Fortun, Kim, James Adams, Tim Schütz, and Scott Gabriel Knowles. 2021. "Knowledge Infrastructure and Research Agendas for Quotidian Anthropocenes: Critical Localism with Planetary Scope." *Anthropocene Review* 8 (2): 169–82. https://doi.org/10.1177/20530196211031972.

Fricker, Miranda. 2017. "Evolving Concepts of Epistemic Injustice." In *The Routledge Handbook of Epistemic Injustice*, edited by Ian James Kidd, José Medina, and Gaile Pohlhaus Jr., 53–60. London: Routledge.

Garba, Tapji, and Sari-Maria Sorentino. 2020. "Slavery Is a Metaphor: A Critical Commentary on Eve Tuck and K. Wayne's 'Decolonization Is Not a Metaphor.'" *Antipode* 52 (3): 764–82.

Gikandi, Simon, and Evan Mwangi. 2007. *The Columbia Guide to East African Literature in English since 1945*. New York: Columbia University Press.

Guevara, Perry, and Amy Wong. 2022. "Grounding the Humanities." *Public Books*, January 20. https://www.publicbooks.org/grounding -the-humanities.

Haraway, Donna J. 2016. *Staying with the Trouble*. Durham, NC: Duke University Press.

Hazelkorn, Ellen, and Georgiana Mihut, eds. 2021. *Research Handbook on University Rankings: Theory, Methodology, Influence and Impact*. London: Edward Elgar.

Herzig, Rebecca. 2006. *Suffering for Science: Reason and Sacrifice in Modern America*. New Brunswick, NJ: Rutgers University Press.

Hountondji, Paulin. 1990. "Scientific Dependence in Africa Today." *Research in African Literatures* 21 (3): 5–15.

Hountondji, Paulin, ed. 1997. *Endogenous Knowledge: Research Trails*. Dakar: CODESRIA.

Hountondji, Paulin. 2009. "Knowledge of Africa, Knowledge by Africans: Two Perspectives on African Studies." *RCCS Annual Review* 1: 121–31.

ICEF Monitor. 2014. "Market Snapshot: Ecuador." June 12. https://monitor .icef.com/2014/06/market-snapshot-ecuador/.

ICEF Monitor. 2015. "New OECD Report Summarises Global Mobility Trends." November 30. https://monitor.icef.com/2015/11/new-oecd -report-summarises-global-mobility-trends/.

Kamau, Gachunga J. 2005. "The Effect of Structural Adjustment Programmes on the University Education Sector in Kenya." MA diss., University of Nairobi. http://erepository.uonbi.ac.ke/handle/11295/20394.

Kawa, Nicholas C., José A. Clavijo Michelangeli, Jessica L. Clark, Daniel Ginsberg, and Christopher McCarty. 2019. "The Social Network of US Academic Anthropology and Its Inequalities." *American Anthropologist* 121 (1): 14–29.

Kervran, David Dumoulin, Mina Kleiche-Dray, and Mathieu Quet. 2018. "Going South: How STS Could Think Science in and with the South?"

Tapuya: Latin American Science, Technology and Society 1 (1): 280–305. https://doi.org/10.1080/25729861.2018.1550186.

Knöchelmann, Marcel. 2019. "A Groundbreaking DEAL?" *Elephant in the Lab*, January 23. https://elephantinthelab.org/a-groundbreaking-deal/.

Knöchelmann, Marcel. 2021. "The Democratisation Myth: Open Access and the Solidification of Epistemic Injustices." *Science and Technology Studies* 34 (2). https://orcid.org/0000-0003-1050-1303.

Kreimer, Pablo, and Hebe Vessuri. 2018. "Latin American Science, Technology, and Society: A Historical and Reflexive Approach." *Tapuya: Latin American Science, Technology and Society* 1 (1): 17–37. https://doi.org/10.1080/25729861.2017.1368622.

Larivière, Vincent, Stefanie Haustein, and Philippe Mongeon. 2015. "The Oligopoly of Academic Publishers in the Digital Era." *PLoS One* 10 (6): e0127502. https://doi.org/10.1371/journal.pone.0127502.

Luepnitz, Deborah. 2009. "Thinking the Space between Winnicott and Lacan." *International Journal of Psychoanalysis* 90: 957–81.

McKittrick, Katherine. 2021. "Public Thinker: Katherine McKittrick on Black Methodologies and Other Ways of Being." *Public Books*, February 1. https://www.publicbooks.org/public-thinker-katherine-mckittrick-on-black-methodologies-and-other-ways-of-being/.

Mignolo, Walter. 2012. *Local Histories/Global Designs: Coloniality, Subaltern Knowledges, and Border Thinking.* Princeton, NJ: Princeton University Press.

Moore, Samuel. 2021. "Open Access, Plan S and 'Radically Liberatory' Forms of Academic Freedom." *Development and Change* 52: 1513–25. https://doi.org/10.1111/dech.12640.

Musila, Grace A. 2019. "Against Collaboration—or the Native Who Wanders Off." *Journal of African Cultural Studies* 31 (3): 286–93.

Muthamilarasan, Mehanathan, and Manoj Prasad. 2014. "Impact of Impact Factor in Quantifying the Quality of Scientific Research." *Current Science* 107 (8): 1233–34.

Muyia, Nafukho. 1996. "Structural Adjustment Programmes and the Emergence of Entrepreneurial Activities among MOI University Students." *Journal of Eastern African Research and Development* 26: 79–90.

Ndlovu-Gatsheni, Sabelo. 2015. "Decoloniality as the Future of Africa." *History Compass* 13 (10): 485–96.

Ndlovu-Gatsheni, Sabelo J. 2021. "The Cognitive Empire and Gladiator Scholarship." *Kujenga Amani* (blog), June 25. https://kujenga-amani.ssrc.org/2021/06/25/the-cognitive-empire-and-gladiatory-scholarship/.

Nelson, Alondra. 2022. "Ensuring Free, Immediate, and Equitable Access to Federally Funded Research." Executive Office of the President. Office of Science and Technology Policy, Washington, DC, August 25. https://www.whitehouse.gov/wp-content/uploads/2022/08/08-2022-OSTP-Public-Access-Memo.pdf.

Newman, John Henry. 1982 [1852]. *The Idea of a University*. Edited by Martin J. Svaglic. Notre Dame, IN: University of Notre Dame Press. https://doi.org/10.2307/j.ctvpj73nm.

Ngũgĩ, Mukoma wa. 2018. "What Decolonizing the Mind Means Today." *Literary Hub*, March 23. https://lithub.com/mukoma-wa-ngugi-what-decolonizing-the-mind-means-today/.

Nkoudou, Thomas Herve Mboa. 2020. "Epistemic Alienation in African Scholarly Communications: Open Access as a Pharmakon." In *Reassembling Scholarly Communications: Histories, Infrastructures, and Global Politics of Open Access*, edited by Martin Paul Eve and Jonathan Gray, 25–40. Cambridge, MA: MIT Press.

Ogone, James Odhiambo. 2017. "Epistemic Injustice: African Knowledge and Scholarship in the Global Context." In *Postcolonial Justice*, edited by Anke Bartels, Lars Eckstein, Nicole Waller, and Dirk Wiemann, 17–36. Leiden: Brill. https://doi.org/10.1163/9789004335196_004.

Okune, Angela. 2020. "Open Ethnographic Archiving as Feminist, Decolonizing Practice." *Catalyst: Feminism, Theory, Technoscience* 6 (2): 1–24. https://doi.org/10.28968/cftt.v6i2.33041.

Okune, Angela. 2021. "Postcolonial Objectivity: Reaching for Decolonial Knowledge Making in Nairobi." PhD diss., UC Irvine. https://escholarship.org/uc/item/7x695678.

Okune, Angela, Sulaiman Adebowale, Eve Gray, Angela Mumo, and Ruth Oniang'o. 2021. "Conceptualizing, Financing and Infrastructuring: Perspectives on Open Access in and from Africa." *Development and Change* 52 (2): 359–72. https://doi.org/10.1111/dech.12632.

Okune, Angela, and Leonida Mutuku. 2023. "Becoming an African Techpreneur: Geopolitics of Investments in 'Local' Kenyan Entrepreneurship." *Engaging Science, Technology, and Society* 9 (1): 81–103.

Piron, Florence. 2018. "Postcolonial Open Access." In *Open Divide: Critical Studies on Open Access*, edited by Ulrich Herb and Joachim Schöpfel, 117–28. Sacramento, CA: Library Juice Press.

Pitts, Andrea J. 2017. "Decolonial Praxis and Epistemic Injustice." In *The Routledge Handbook of Epistemic Injustice*, edited by Ian James Kidd, José Medina, and Gaile Pohlhaus Jr., 149–57. London: Routledge.

Posada, Alejandro, and George Chen. 2017. "Publishers Increasingly in Control of Scholarly Infrastructure and This Is Why We Should Care." *The Knowledge G.A.P.* (blog), September 20. https://worldpece.org/node/1141.

RELX. 2019. "Results for the Year to December 2018" (Press release). London: RELX Group, February 21. https://www.relx.com/~/media/Files/R/RELX-Group/documents/press-releases/2019/relx-results-2018-pressrelease.pdf.

Rivera Cusicanqui, Silvia. 2012. "Ch'ixinakax utxiwa: A Reflection on the Practices and Discourses of Decolonization." *South Atlantic Quarterly* 111 (1): 95–109.

Rodriguez Medina, Leandro. 2018. "Internationalizing Science and Tech-
    nology: Some Introductory Remarks." *Tapuya: Latin American Science,
    Technology and Society* 1 (1): 216–18. https://doi.org/10.1080/25729861
    .2018.1550968.

Sariola, Salla. 2021. "Editorial—What Does Openness Conceal?" *Science and
    Technology Studies* 34 (2): 2–5. https://doi.org/10.23987/sts.107722.

Scott, Joan W. 2019. *Knowledge, Power, and Academic Freedom.* New York:
    Columbia University Press.

Sekhsaria, Pankaj. 2019. *Instrumental Lives: An Intimate Biography of an
    Indian Laboratory.* London: Routledge.

Smith, Linda Tuhiwai. 2014. *Decolonizing Methodologies: Research and Indig-
    enous Peoples.* London: Zed.

Solano, Gonzala. 2020. "Ecuador Announces Major Cuts in Public
    Spending." Associated Press, May 19. https://apnews.com/article
    /3179a21a1b94b21cb9991856fa575fc4.

Sørenson, Knut H., and Sharon Traweek. 2021. *Questing Excellence in Aca-
    demia: A Tale of Two Universities.* London: Routledge.

Soto Laveaga, Gabriela. 2020. "Moving from, and beyond, Invented Catego-
    ries: Afterwords." *History and Theory* 59 (3): 439–47.

Suarez, Maka, and Jorge Núñez. 2020. "Social Mobilizations in Ecuador:
    From October 19 to Covid-19." *Member Voices, Fieldsights*, November 24.
    https://culanth.org/fieldsights/social-mobilizations-in-ecuador-from
    -october-19-to-covid-19-1.

Thiong'o, Ngũgĩ wa. 2011 [1986]. *Decolonising the Mind: The Politics of Lan-
    guage in African Literature.* Nairobi: Heinemann Educational.

Thondhlana, Juliet, Evelyn Chiyevo Garwe, Hans de Wit, Jocelyne Gacel-
    Ávila, Futao Huang, and Wondwosen Tamrat, eds. 2020. *The Bloomsbury
    Handbook of the Internationalization of Higher Education in the Global
    South.* London: Bloomsbury.

Torres, Wilmer. 2020. "Universidades: Recorte alcanza a las privadas
    cofinanciadas y de posgrado." *Primicias*, May 6. https://www.primicias
    .ec/noticias/sociedad/recorte-presupuestario-universidades-privadas
    -posgrado/.

Tousignant, Noémi. 2018. *Edges of Exposure: Toxicology and the Problem of
    Capacity in Postcolonial Senegal.* Durham, NC: Duke University Press.

Traweek, Sharon, Duygu Kaşdoğan, and Kim Fortun. 2021. "Searching for
    How Epistemic Power Is Made, Appropriated, Circulated, and Chal-
    lenged: An Interview with 2020 4S Bernal Prize Winner Sharon Traweek."
    Engaging Science, Technology, and Society 7 (2): 97–119. https://doi.org/10
    .17351/ests2021.1247.

Tsing, Anna L., Heather A. Swanson, Elaine Gan, and Nills Bubandt, eds.
    2017. *Arts of Living on a Damaged Planet: Ghosts and Monsters of the An-
    thropocene.* Minneapolis: University of Minnesota Press.

Tuck, Eve, and Wayne K. Yang. 2012. "Decolonization Is Not a Metaphor."
    Decolonization: Indigeneity, Education and Society 1 (1): 1–40.

Ureta, Sebastian. 2021. "Ruination Science: Producing Knowledge from a Toxic World." *Science, Technology, and Human Values* 46 (1): 29–52. https://doi.org/10.1177/0162243919900957.

Vaughan, Tyson. 2021. "Re-mooring: Rethinking Recovery and Resilience in the Anthropocene." In *Disastrous Times: Beyond Environmental Crisis in Urbanizing Asia*, edited by Eli Elionoff and Tyson Vaughan, 196–214. Philadelphia: University of Pennsylvania Press.

Vessuri, Hebe. 2019. "Crises That Mismatch Canons in Science: Provincialization, Transnationality, Conviviality?" *Tapuya: Latin American Science, Technology and Society* 2 (1): 26–31. https://doi.org/10.1080/25729861.2019.1586193.

Wilder, Craig Steven. 2013. *Ebony and Ivy: Race, Slavery, and the Troubled History of American Universities*. London: Bloomsbury.

Agroecological Innovation

Decentralizing Knowledge and Democratizing Brazil's Agrifood Economy

Modernist agendas to centralize and standardize knowledge have facilitated the dual exploitation of people and nature, while instrumentally framing them as labor and natural resources, respectively. Neocolonial regimes have centralized extractivist strategies to plunder the land, especially through minerals and agro-industrial primary products for export. Claims to modernize production through technoscientific knowledge have sanitized this plunder as efficient, enlightened progress. It appropriates, marginalizes, or devalues other knowledges, meanwhile obscuring their epistemic contribution. This capitalist modernity extends an earlier coloniality imposing exploitative relationships; in response, subaltern voices and practices often contest them (Mignolo 2000).

As Linda Martín Alcoff notes: Resource-extraction agendas transform economies, political-legal institutions, and the organization of labor, thus rupturing earlier social identities and communal relations. An extractivist epistemology treats its epistemic resource as separable from its origin. Moreover, even knowledge can become a commodity with exchange value, whose exclusive rights can be contractually defined, protected, and enforced (Alcoff 2022, 213; reprinted in this volume as chapter 1).

Epistemic extractivism describes how modern science has appropriated knowledge and erased the historical memory of its noncapitalist sources, while also preempting alternative futures (Grosfoguel 2019). Such plunder has often marginalized subaltern social groups or even eliminated them. This has destroyed the knowledge of peasant, indigenous, and Afro-descendant groups. "This epistemicide is the death of peoples. The university owes an historical debt for having produced this epistemicide" (de Sousa Santos 2009, 123).

As a further stage since the 1980s, the old concept of efficiency has become debased by serving external values of neoliberal globalization rather than local needs. Moreover, "the discursive hegemonic model of neoliberalism, protected by modernization and efficiency, has demanded that the university adapt to it" (de Sousa Santos 2009, 94). Consequently, the university has been a focal point of struggles over knowledge politics.

For contesting modern science as a monoculture, a key concept has been *ecología de saberes*, ecology of knowledge. This promotes dynamic interactions among plural, heterogeneous knowledges, without undermining their autonomy (de Sousa Santos 2007, 85). Articulating diverse knowledges helps to imagine alternative forms of awareness and life (de Sousa Santos 2009, 14).

A new opportunity came with Bolivia's 2009 Constitution for a plurinational state, recognizing indigenous peasant cultures and communitarian economies. The Constitution provided for decentralizing the nation-state through the regional departments, which could create more favorable conditions for research agendas to accommodate local demands (de Sousa Santos 2009, 82–83). As it turned out, this happened only in departments whose governments accommodated peasant movements, while others promoted neocolonial agendas with their hegemonic knowledges.

Alongside its extractivist agenda, epistemic centralization has provoked widespread resistance, counterhegemonic concepts, and alternative agendas. As one form, epistemic decentralization expresses efforts to shift epistemic agency from the centered networks, institutions, and technologies to the peripheries. This often happens through socioenvironmental conflicts and/or socio-technical controversies. They decentralize local knowledge, activate citizens and governments, and contribute to the production of sociopolitical changes that generate decentralized agendas linking local priorities with global agendas (introduction, this volume).

Through such efforts, the global has been given counterhegemonic meanings, linking social movements transnationally. The center/periphery binary originally highlighted neocolonial relationships between the Global North and South. That binary has increasingly described neocolonial relationships within countries, as their governments have extended or replicated northern extractivist agendas in their own country. Decolonial responses have linked various resistances within and across countries.

Neocolonial epistemic politics have been contested and theorized, especially in Latin America (as in citations above). Counterhegemonic concepts emerged there over several decades from networks of social movements, practitioners, and academics; they jointly contributed to Latin American critical thought. This has critically analyzed the Eurocentric neocolonial meanings and roles of hegemonic concepts, such as modernity, modernization, technology, innovation, territorial development, nature protection, and so on. To counter their role, decolonial counterhegemonic concepts have been elaborated (e.g., Haesbaert 2021; Lander 2001; Mignolo and Walsh 2008).

Latin American critical thought links three perspectives: the close relationships between humans and the land they inhabit, countering the terra nullius justification for colonialism; the prime importance of traditional knowledge for current alternatives; and the relationship between agroecological practices and decolonial struggle (Rosset et al. 2021). Indeed, an agroecology-based solidarity economy has devised counterhegemonic agendas for decentralizing and democratizing knowledge. Such efforts have influenced some government policies, state agencies, and their support measures, with wider lessons for institutional strategies, as this chapter demonstrates for Brazil.

This chapter deals with three research questions: (1) How has the neocolonial modernization project, along with its epistemic centralization, encountered popular resistance, especially from solidarity economy networks? (2) How have they counterposed alternative agendas and gained some influence within institutions? (3) Toward an agroecology-based solidarity economy, how does agroecological innovation attempt to democratize the economy and decentralize knowledge?

Here those questions will be answered for Brazil in its Latin American context. This chapter has the following structure. The first section discusses Latin America's solidarity economy agenda as a means to link economic democratization and epistemic decentralization, especially from a decolonial perspective. The second section covers rival agrifood agendas: agroindustrial versus agroecological, especially in Brazil. The third section describes Brazil's convergence between *economia solidaria* and agroecological innovation, linking democratization with decentralization. The conclusion answers the research questions.

The chapter emphasizes Latin American concepts, especially from Brazil. The author searched for the terms *epistemic* and *decentralization*, which have a rich history over the past two decades. These concepts have provided a basis for EcoSol-agroecology agendas. Quotes and paraphrases should help readers to understand the connection. All those sources have been translated by the author.

Economia Solidaria (EcoSol): Decentralizing Knowledge

Solidarity economy movements have promoted a worldwide agenda to transform economic activity and knowledge along democratic lines (Utting 2018). These movements have promoted mutual aid activities seeking to improve livelihoods, while bypassing profit-driven middlemen and markets. This agenda develops "a framework for *horizontal* relations between persons and social collectives in their quest to satisfy their common needs." The agenda emphasizes women's leadership toward overcoming women's socioeconomic inequalities, especially in the informal economy (RIPESS 2015, 2–5; see also RIPESS 2012).

This agenda faces major obstacles, especially the *homo economicus* model, Cartesian dichotomy, and logical-positivist epistemology. All these constrain our abilities to understand economic alternatives based on other rationalities. Hence the need for an epistemological transformation alongside a practical one. This must go beyond the stereotypical binary of rational/affective capacities (Dash 2014, 2016), and likewise beyond the "iron cage" of instrumental rationality (Leff 2009). This denotes instrumental decision-making based on rational calculation of what would be materially beneficial. The term *iron cage* comes from Talcott Parsons's loose translation of "a shell as hard as steel" from Max Weber's (1930) book, *The Protestant Ethic and the Spirit of Capitalism*.

In Latin America the solidarity economy is called Economia Social y Solidária (ESS), or often EcoSol for short. The EcoSol agenda has built cooperative relationships among economic activities, while gaining collective capacities for democratic self-management (dos Santos and Carneiro 2008; Schüttz and Gaiger 2006; Singer and Souza 2000). This alternative also depends on mobilizations demanding better public services, support measures, and labor protections (Coraggio 2016, 24).

Short supply chains, often called *circuitos curtos*, bring producers closer to consumers while avoiding profit-driven middlemen. The term *curto* (short) denotes the social proximity of mutual aid and reciprocity. This agenda has given practical meaning to calls for epistemic decentralization through alternatives to the dominant capital-intensive modernization model of development.

Early on, Brazil's EcoSol movement contributed to the wider movement against the military junta of 1964–85. Looking beyond the junta, the movement sought to ensure a continuous democratization process, seeing decentralization as a crucial means. This aim was somewhat accommodated by a state commitment to multilevel democratization. The postjunta Constitution affirmed "political-economic decentralization, encompassing the coordination and general norms in the federal sphere and the coordination and execution of the respective programs in the state and municipal spheres, as well as social assistance charities" (Brazil 1988, Article 204).

Democratizing social assistance would accommodate criticisms of the prevalent *assistencialista* approach, which perpetuated the needy condition of the people receiving assistance. Critics counterposed a solidaristic approach helping beneficiaries to gain autonomy to follow their own pathway. As a turning point, the 1993 Lei Orgânica de Assistência Social (LOAS) reformed social assistance according to the Constitution's principles of democratization and decentralization. The challenge to establish democratic, horizontal forms was taken up by the network of public policy managers advocating a solidarity economy (Rede de Gestores 2008).

Knowledge exchange among diverse cultures has been crucial for a solidarity economy: "To elaborate proposals that contribute to an EcoSol public politics requires us to observe the diverse complexity of experiments that express and, at the same time, represent means to cultivate and make enjoyable the diverse forms of production and its understanding, based on diverse cultural patterns, which otherwise define the necessities, abandon the materialistic and instrumental culture, and establish other relationships between human beings and nature. The alternative sources of knowledge are stimulated by alternative sources of production" (de Sousa Santos and Rodriguez 2002).

From a critical epistemological perspective, the concept of social management points toward management organization from the bottom up through horizontal decision-making (Monje-Reyes 2011, 705). "Examining the epistemological bases of the solidarity economy and its articulating conceptual link to socio-politico-economic decentralization, its conceptual link is workers' participation in their own decisions about production and distribution of its results. Therefore it is a form of productive organization highly linked with participatory and protagonistic democracy of the social subjects " (Monje-Reyes 2011, 720–21).

Influenced by indigenous and subaltern cultures, this Latin American epistemology is often called *sentipensante*, combining the heart (*corazón*) with the thought process (Fals-Borda 2008). Here the heart constitutes the ontological core of a worldview proper to Latin American epistemologies. Those aspects have been conceptualized as the epistemic paradigm of the countryside (Barbosa 2016), involving defense of territories and of Tierra Madre, in both a communal and agroecological sense (Barbosa 2019). This cultivates agrobiodiversity together with ethno-cultural diversity (Leff 2001). This inner connection is conceptualized as agro-socio-biodiversity, rich knowledge systems that are maintained by farmers' everyday social role in generating and managing biodiversity; this helps to resist its degradation by agribusiness (Schmitt 2018, 2, 10).

The EcoSol-agroecology convergence has been informed likewise by the Andean concept of *vivir bien*: that humans should live in harmony with each other and with Tierra Madre, respecting her rights. The concept originates from indigenous Andean languages (Bolivia 2009). It counters reductionist binaries, such as nature/culture and subject/object, which pervade the European/northern thought underlying the dominant agrifood system (Rosset et al. 2021, 638–39). *Vivir bien* promotes an ideal future, linking and inspiring current practices that resist the dominant forces threatening such harmony.

Rival Agrifood Models

The above perspectives link economic democracy and epistemic decentralization of producers for a decolonial empowerment vis-à-vis the dominant agrifood system. All this provides a deep cultural basis for EcoSol-agroecology agendas in Latin America, amid conflicts between rival agrifood systems. This section sketches how the dominant model has provoked resistance and such alternatives there, especially in Brazil.

Contested Agro-Industrial Food System:
Inequitable Modernization

As the global context for such conflicts, the agricultural modernization agenda has fundamentally transformed farmers' relationships with nature, land, and consumers. It has shifted farmers' practices toward more complex technological tools and institutional arrangements seeking a competitive advantage in distant markets. State and corporate structures have centrally planned a societal change that conceives farmers and their communities as passive receptacles for new ways to do agriculture—rather than as social actors bringing their own projects, capacities, and trajectories (Long and van der Ploeg 2011; van der Ploeg and van Dijk 1995).

The dominant system structures production for a competitive advantage in distant anonymous markets. It combines capital-intensive innovation, a techno-diffusionist centralized model of expert knowledge, resource plunder, and globalized supply chains for standard commodity crops. This system has undermined producers' livelihoods, exhausted natural resources, and degraded food quality via ultra-processing, especially in the Global South.

Whenever small-scale producers have attempted to imitate or accommodate this model, male farmers have more easily accessed loans to buy technology packages and then awaited payments for the harvest. But they face structural disadvantages in competing on the same terms; they lose much of the value added to profit-driven middlemen. Some have incurred long-term debt and abandoned farming.

Throughout Latin America in particular, agro-industrial systems have led farmers into capital-intensive monocultures and agrochemical inputs, degrading the natural resource base. In many places, such systems have undermined small-scale producers, their livelihoods, traditional knowledge, land rights, and locally developed biodiverse seeds (PIADAL 2013). Despite this damage, Latin American governments have generally promoted such systems.

Brazil's agri-modernization agenda has had several stages. During 1964–85, the military junta promoted technology packages and loans benefiting mainly larger-scale farmers to increase productivity for export markets; this so-called conservative modernization constrained the agrarian reform objectives to broaden land tenure. In the early 1990s, the postjunta government identified 2.4 million agricultural units able to achieve a "family farm transition" to modern agriculture by imitating large-scale agro-industrial methods. It categorized approximately half the family farms as "transition family farming," that is, conducive to modernization (Guanziroli, Buainain, and Sabbato 2013).

From 1995 onward, Brazil's PRONAF (National Program for Strengthening Family Farming) funded and encouraged such peasants to imitate the capital-intensive methods of large farms. Through chemical-intensive technology packages, PRONAF meant to increase productivity, alleviate poverty, and enhance food security. This techno-diffusionist development model has dominated rural extension services. Agricultural knowledge always comes from producers, yet techno-diffusionism reserves this capacity to a few experts, thus wasting the cognitive capacities inherent in everyone. This resulted in technological dependence, which "translates into cultural dependence, immobilizing local autonomous innovation capacities, thus diminishing the manoeuvre-room of rural families and communities for self-determination through a permanent implementation of their technical-economic strategies" (ANA 2007, 7).

Benefiting from those policies, Brazil's large agro-industrial chemical-intensive farms hold most of the land but produce mainly export commodities (IBGE 2006, 2017). By contrast, Brazil's 4.4 million family farms comprise 85 percent of agricultural establishments; they use less than 25 percent of the agricultural land to produce 70 percent of the food consumed nationally. Some have adopted capital-intensive methods similar to agribusiness, yet most still continue traditional practices. Such smallholders still produce most of the food that is sold by wholesalers through supermarket chains and outdoor markets; those middlemen keep consumers separate from producers, who lose much of the sale price.

This outcome has been analyzed as a "new unequal modernization," aggravating social exclusion and regional inequalities (Tonneau, de Aquino, and Teixeira 2005). Most peasants have faced structural constraints in gaining the promised benefits. Hence various studies have recommended more appropriate support measures for the majority of farmers who cannot or will not follow the capital-intensive agri-modernization pathway (Cabral et al. 2016; Medina et al. 2015).

Agroecological Alternatives as Epistemic Decentralization in Latin America

Given the many harms from the hegemonic system, more peasants have been rejecting the prevalent language of modernization, efficiency, productivity, economies of scale, trade liberalization, free markets, and so on. They seek relative autonomy from competitive markets for credit, inputs, and outputs (Rosset and Martínez-Torres 2012, 17). Such movements have opposed a rural development model that seeks to industrialize agriculture and thus "modernise the countryside to bring it out of backwardness," as Caporal and Costabeber (2004, 6) put it sarcastically.

As a significant alternative to the hegemonic system, preindustrial cultivation methods have been reconceptualized and popularized as agroecological. These have been improved through agro-innovations in-

Figure 7.1 "Ecology of Knowledge: Science, Culture and Art," poster, 2019. The illustration shows how agroecology movements seek to democratize knowledge and agrifood systems. Source: XI Congresso Brasileiro de Agroecologia, Universidade Federal de Sergipe (UFS).

tegrating traditional and modern scientific knowledge of agroecosystems, which depend on biodiverse ecological relationships within and beyond a farm. As an organizational innovation, short supply chains have built closer relations between small-scale agroecological producers and consumers, who thereby support the production methods.

These agroecological initiatives exemplify territorial markets that specialize in selling food that is produced, distributed, and consumed within a specific territory (Kay 2016). Such arrangements that serve local food needs also strengthen livelihoods of local food producers, processors, and vendors. Meanwhile, they reduce participants' dependence on transnational corporations that dominate and concentrate global supply chains (HLPE 2020). Policy support measures are necessary for smallholders to access and create congenial markets, so that traditional methods can gain better access to the food value chain, argues the FAO's High-Level Expert Panel (HLPE 2013; 2019, 110).

Agroecology agendas reframe those issues. Their narratives motivate farm families to improve their traditional cultivation methods, or else to undertake difficult shifts away from agro-industrial methods, both understood as an agroecological transformation. Knowledge-intensive artisanal agroecological methods use locally available resources, toward an agriculture that is socioenvironmentally and economically sustainable (Caporal and Costabeber 2004, 79). This needs strategies that decentralize productive processes so that they are "compatible with the ecological conditions and capacities to incorporate ethnic identities and their respective cultural values" (83).

Such methods reproduce biodiverse seeds, maintain wider biodiversity for crop resilience, and recycle nutrients, together minimizing environmental harm. These practices likewise enhance livelihoods by avoiding external inputs and so minimizing costs. This continent-wide agenda eventually became grounded in social agrarian agendas and social movements, especially in resistance to capitalist modernization, that is, dependence on technology packages of the Green Revolution (Altieri 2002; Altieri and Nicholls 2008; Altieri and Toledo 2011).

Such efforts by peasant and civil society movements have been informed by an epistemic decentralization. While such agrofood alternatives have arisen globally, Latin American ones have many special characteristics. First, an epistemic understanding of territory confers an identitarian ethos in the political narrative of indigenous and peasant movements; they defend land, nature, and common goods, recognizing territory as a space for the reproduction of life. Second, a biocultural memory of traditional knowledges often underlies efforts to preserve native seeds, conserve soils, make sustainable use of water, and recover ancestral agroecological practices. Third, rural social movements deploy agroecology as a banner of collective struggle, cultural (re)construction, and a political project defending its territories for food sovereignty (Rosset et al. 2021).

Brazil's Extension Services: Epistemic-Organizational Shift

In the 1980s a distinctive agroecology movement was first arising, organized around grassroots groups called Comunidades Eclesiais de Base (CEBs). To obtain appropriate advice, they sought changes in the agricultural extension services. The encounter between them and CEBs "was done through a real epistemological shock" as regards their political commitment to the peasantry and to its popular wisdom for local development. At that time, technical advisors were mostly professionals who had been trained academically in techniques for expanding capitalist agri-production. They had difficulties in detaching themselves from the productivist model. And they lacked the conceptual tools to understand ecological aspects of family farming (ANA 2007, 11).

Jumping ahead a couple decades: After the 2003 general election, a significant shift came from the government led by the Partido dos Trabalhadores (PT). In response to social movements, the Brazilian state gradually established or shifted technical assistance services to support agroecological practices. Since 2004, the Programa Nacional de Assistência Técnica e Extensão Rural (PRONATER or PNATER) has promoted farm-level experiments with technologies more appropriate for smallholders, including agroecological methods (Schmitt et al. 2017, 88–89). Even for such methods, however, agri-extension services largely continued the techno-diffusionist approach for various reasons, for example, because some internal forces sought to maintain it, or because staff formalized agroecological methods as separate techniques (ANA 2007, 9).

Indeed, when small-scale farmers received subsidies or loans under the Programa Mais Alimentos, they bought machinery and equipment, stimulating them to cultivate larger areas with more external inputs (Mussoi 2011, 182). Agroecology was often adopted as a simple input substitute for technology packages of the Green Revolution (256). When providing support measures to small-scale farmers, often the state research agency Empresa Brasileira de Pesquisa Agropecuária (Embrapa) promoted a portfolio of agroecology technologies, thus perpetuating the techno-diffusionist approach (Petersen, Mussoi, and Dal Soglio 2013, 110, 112). This likewise extended a centralized vertical regime of professional experts, now in greater conflict with agroecology support networks.

Analogous tensions arose within Brazil's agri-research institute, Embrapa. It extended the dominant techno-diffusionist perspective from the Global North, as a basis for independent agro-innovations that would help agriculture to increase productivity and so gain export markets. Although dedicated mainly to research and development, a techno-diffusionist perspective became more explicit and active in 2009 through a new Technology Transfer Directorate. This set up decentralized units but still within a vertical techno-diffusionist perspective. There ensued a long debate over whether Embrapa's role should be technology transfer or knowledge sharing; the latter meant a horizontal process involving agri-extensionists and farmers who were generating innovation (Borsatto, Bergamasco, and Bianchini 2017). A horizontal knowledge-sharing approach emerged, especially from Embrapa's agroecology research units, which were being established or expanded during the first PT-led government (Borsatto, Bergamasco, and Bianchini 2017; ANA 2007, 150).

For both research and extension services, a cultural shift happened by several means. The Articulação Nacional de Agroecologia (ANA) investigated and facilitated a decentralization process, in both the epistemic and organizational senses. For its first major report on the shift and its difficulties, the investigation process reflected on the concrete practices of agroecological advice. On a national scale, this was being decentralized by some advisory agencies, involving farmers' organizations whose members received technical advice. Agencies made such improvements through diverse pathways (ANA 2007, 15).

As a key impetus and capacity for such a change, civil society groups had long criticized the hegemonic diffusionist model and developed horizontal alternatives (ANA 2007, 9). Drawing on the concept of *ecologia de saberes* (de Sousa Santos 2007), they promoted *diálogo de saberes*, toward a "dialogue capacity and collective learning." This has developed a peasant-based expertise, reinforcing their demands for rural extension services that do likewise. This enhances sustainability by improving everyday practices, rather than by providing external techniques (Caporal and Costabeber 2004, 120; see also Delgado and Rist 2016; Holt-Giménez 2006).

An agroecological peasant-based perspective made more visible the cultural heritages of various regions, as seen in diverse food customs, cultivars, and herbal home remedies based on them (ANA 2007, 133). In some places, food-processing units were being decentralized by farmers' cooperatives. This facilitated the greater use of the rich biodiversity of native plants (ANA 2007, 149).

Such biodiversity soon encompassed Plantas Alimentícias Não Convencionais (PANCs, nonconventional food plants), which valorize agrobiodiversity and its related sociocultural diversity (GVC 2015). The PANCs provide means to produce herbal medicines and/or traditional foods. More generally, fruits are lightly processed into tasty products that have a longer shelf life and gain more income. This avoids waste, which befalls approximately 40 percent of fresh food in Brazil.

Brazil's EcoSol-Agroecology Agenda as Collective Capacities

Since the 1990s, Brazil's peasant and civil society movements have been demanding state measures to support traditional farming skills, biodiverse resources, fairer market access, and collective capacities for cooperative organizations. Over the past decade, social movements previously promoting either EcoSol or agroecology converged toward integrating them. Women have played central roles in both social movements, as well as in their convergence; Brazil exemplifies women's continent-wide role (SOF

2015; Zuluaga et al. 2018). With decolonial feminist perspectives, they have contested the epistemic basis of the dominant agri-technoscience and brought epistemological innovation (Zuluaga et al. 2018). These efforts both depend on and build support networks for collective capacities, as shown in this section.

EcoSol-Agroecology Convergence

The ANA has brought together diverse social movements, nongovernmental agencies (NGOs), and policy managers promoting agroecological alternatives to the dominant agrifood system. As a turning point, in 2011 the ANA co-organized an event to merge the agendas for agroecology and the solidarity economy (FBES 2011; Schmitt 2020, 39). As a key role, agroecology could help conserve natural and cultural heritages, as well as build local networks of a solidarity economy (ANA 2012, 3). Conversely, agroecology was being incorporated into the EcoSol agenda (FBES 2012).

That convergence of social movements was accommodated by the 2003–16 governments led by the PT. After a decade, its 2013 Plano Nacional de Agroecologia e Produção Orgânica (PLANAPO) brought together many relevant policies, including solidarity economy (CIAPO 2013, 16). Conversely, its solidarity economy policy emphasized such a basis for technical training in agroecology (CNES 2015, 16, 32).

For example, the Fome Zero (Zero Hunger) campaign was structured in ways that created new markets for small-scale peasant farmers. Secure land tenure was granted to many more land occupations, whose settlers began to adopt agroecological methods. Agroecological knowledge exchange has helped peasants to better use locally available natural resources rather than external inputs, thus countering the dominant agrimodernization process (van der Ploeg, Ye, and Schneider 2012, 134, 147).

Toward an agrifood solidarity economy, in the 1990s several civil society organizations began promoting organic certification through the Sistema Participativo de Garantia (SPG). These systems remain under producers' control, rather than relying on expensive third-party *auditoria* systems (Schmitt et al. 2017, 85; Schwab and Collado 2017, 2). The PT government eventually accommodated such proposals through a new model, Organização de Controle Social (OCS): each cooperative has "a relationship of organization, commitment and trust amongst the participants"; this system aims at "stimulating a direct relationship between the producer and final consumer" (MAPA 2007, 3; see also MAPA 2008). Alongside the lower

Figure 7.2 Logo of Fórum Brasileiro de Economia Solidária (FBES), illustrating Brazil's social movements for *economia solidária*. See https://fbes.org.br/.

Figure 7.3 Logo of Articulação Nacional de Agroecologia (ANA), an organization connecting Brazil's social movements for agroecology with policymakers. See https://agroecologia.org.br/.

cost, peasants have preferred this system as less bureaucratic than professional third-party certification.

In each initiative, small-scale producers cooperatively build the OCS-SPG system through their own local norms. Solidaristic participation of all stakeholders ensures the quality of the production process and the final product. At each stage, documentation of the process builds credibility (Ecovida 2003). Their collective capacities have been built through various NGOs, such as the Assessoria e Serviços a Projetos em Agricultura Alternativa (AS-PTA) and the Rede de Agroecologia Ecovida. Organic certification helps small-scale farmers to gain advantageous institutional markets through state procurement programs.

For capacity building, the Programa de Aquisição de Alimentos (PAA) brought together small-scale family farmers (especially women) to learn cooperative skills for collective marketing. A major opportunity has been public procurement programs, especially the Programa Nacional de Alimentação Escolar (PNAE), which has purchased food for school meals. Under PNAE, public institutions pay a 30 percent premium price for organic and agroecological products, making these methods economically more viable for producers. The program aims to promote the biophysical development, learning, and training in healthy food habits of students, especially to fulfill their nutritional needs during the school term (Brazil 2009). Those local procurement programs have favored agroecological products from small family farms (CIAPO 2013). Such programs have helped farmers to strengthen their self-esteem, improve their agroecological methods, and diversify their production (Grisa et al. 2009).

Originating in the wider EcoSol movement, *social technology* denotes a design and use promoting social aims such as collective capabilities, inclusion, and socioeconomic equity (Dagnino 2010; Fressoli and Dias 2014; ITS 2004; Serafim, de Jesus, and Faria 2013; Pires and Novaes 2016, 116). Artisanal skills are adapted in new ways, rather than replaced by technology (Dagnino 2010). Through social technology, production methods are cheaply developed, consolidated, and appropriated by the producers, as a basis for replicating them elsewhere.

The concept of social technology was popularized through new institutions. With funds from the Banco do Brasil, the Instituto de Tecnologia Social stimulated such innovations in many areas, including agroecology. A new model, the Incubadora de Tecnologia Social (ITS), has promoted production-consumption chains valorizing agrobiodiversity as products of socio-biodiversity, alongside youth participation and more equal gender relations. Agroecological initiatives have taken up these concepts and networks to improve production methods, as well as to limit or remediate pollution from agro-industry.

The above concepts have been further elaborated as socioenvironmental technology, which has been defined along lines of sustainable development, namely techniques facilitating practices that are environmentally sound, socially just, economically viable, culturally acceptable, and easily replicable. Such innovations aim to use locally available resources, exchange knowledge, use appropriate didactic tools, create knowledge multipliers, and spread environmental responsibility (IAM 2016).

Knowledge exchange networks have been broadened to encompass all those sociotechnical aspects, as exemplified by the Programa Ecoforte (ANA, 2019b). "The knowledges associated with these experiences are shared and enriched as a common good and source of mutual learning" (Schmitt 2020, 23). The *sentipensante* epistemology has been extended to socioenvironmental technologies, while also reclaiming territory for resource conservation.

EcoSol-Agroecology Support Networks

At most outdoor food markets, called *feiras livres*, stallholders obtain products through a wholesale intermediary; the products have anonymous sources. By contrast, agroecological initiatives have established *circuitos curtos* (short supply chains) for reaching consumers, on the basis that their purchases support cooperative work organization and environmentally

Figure 7.4 Logo of Ação Coletiva de Comida Verdade (Collective Action for Real Food), a group founded during the COVID-19 pandemic to support provision of *comida de verdade* (real food). See https://acaocoletivacomidadeverdade.wordpress.com/.

sustainable practices. Agroecological farmers have organized distinctive *feiras do agricultor* (farmers' markets), which describe their products under various terms such as *real food* (*comida de verdade*) or *peasant food*, indicating that they were produced without pesticides. Often the municipality provides suitable facilities.

Agroecological farmers also sell their products through food baskets for subscribers of community-supported agriculture. Schemes are organized either by farmers or by "conscientious consumers" promoting consumer education about agroecological methods (Matte and Preiss 2019). Civil society groups such as EcoSol have provided advice on establishing those short supply chains. Small-scale farmers thereby bypass conventional markets, rather than being pushed into a futile competition for low prices per se.

Circuitos curtos involve several functions: "At a territorial scale, the networks take on different functions, which relate to several roles: [to?] managing knowledge; to articulating between production, food processing and commercialization; to strengthening sociocultural identities; to generating credibility for product quality; and to coordinating public policies, among others" (Schmitt 2020, 82). Such networks are "organizations acting in a given territory and interacting through a participatory, cooperative dynamic," broader than formal cooperatives (71). They create synergies across activities, thus enhancing the effects of public policies, while also proposing how to improve them (12–13).

Such solidaristic markets, collective self-certification schemes, and their public credibility have been facilitated by national support networks such as Rede Ecovida. Since 2006, its member-producers have coordinated long-distance transport networks for flexibly moving products from places where they are in surplus, thus maximizing producers' income and consumers' food diversity (van der Ploeg, Ye, and Schneider 2012, 158–59; see also Schmitt 2020). Participants periodically agree on product swaps, their prices and operational costs to be shared (Magnanti 2008).

Bem viver both expresses and inspires capacities to use endogenous resources to strengthen social wealth. Through strategic management of such resources, peasant agriculture establishes a coproduction between nature and social institutions. This helps them to reproduce social values in ways relatively autonomous from markets (Petersen, Mussoi, and Dal Soglio 2013). *Circuitos curtos* build consumer support for nature conservation and culturally diverse foods (Schmitt 2020, 71).

The modernization process has inflicted multiple damages necessitating more remedial care activities, while placing the major burden upon women. In response, feminist initiatives have sought to share, revalorize, and de-domesticate women's roles in social reproduction (Hillenkamp, Guérin, and Verschuur 2014, 18; 2017, 52). Going beyond the false alternatives of unpaid or waged labor, EcoSol activities have made women's care roles more visible and valorized (ActionAid Brasil 2010).

All this has been an impetus for women to join or lead EcoSol-agroecology alternatives. Playing a central role, the Sempreviva Organização Feminista (SOF) is a civil society organization promoting a triple agenda of "social movement, transformation and feminism." It has stimulated or organized many training activities, including technical skills in agroecological production and collective marketing. It situates agroecology within a wider struggle of women for control over their bodies, livelihoods, and territory (Lobo 2021).

According to SOF, "The solidarity economy continues in traditional communities (indigenous, quilombolas and peasant) when their labour and territorial management are organized with respect for everyone and Nature" (Nobre, Faria, and Moreno 2015, 18). Crucial for democratic self-management, "horizontality must be seen as a political commitment and as a process not exempt from contradictions, for example, the group's difficulty in making decisions" amid various specializations of its members (Nobre, Faria, and Moreno 2015, 32). In these ways, feminist approaches seek to overcome class, gender, and ethnic-racial inequalities through alternative economies that build resistance to capitalist, patriarchal domination.

Through women's leadership in EcoSol-agroecology initiatives, they have exercised more decision-making over family income and agroecosystem resources (ActionAid Brasil 2010, 272). Such initiatives have become larger and more diverse, while also contesting gender stereotypes and overcoming inequalities (Schmitt 2020, 277). Sometimes called *protagonismo feminino*, women's leadership has enlivened such initiatives while reconfiguring women's roles in family agriculture (ANA 2019b, 21). The concept of *protagonismo feminino* has inspired women's networks in EcoSol-agroecology networks.

Horizontal nonmonetary resource flows build social reciprocity, social capital, and a capacity for joint activity and care in production processes. All this can protect or build a local culture and thus a communitarian sense of belonging (Schmitt 2020, 273, citing Sabourin 2011). Such practices draw on mutual-aid traditions, known as *mutirão*, which go beyond their traditional social base.

Likewise central to a solidarity economy, the concept of *homo situs* denotes a "recomposed man," who identifies with a place as a symbolic site of belonging. Diverse contributions are integrated into a "composite culture of social networks and belonging" (Zaoual 2010, 31, 34). In this sense, *circuitos curtos* depend on "a geographical and relational proximity between producers and consumers" (Darolt, Lamine, and Brandemburg 2013, 10). As a more comprehensive framework, common aims activate many forms of proximity: organizational, institutional, and cultural (Calgaro, da Silva, and Santos 2022; Levidow, Sansolo, and Schiavinatto 2022; AgroEcos 2022).

This agenda aims to decentralize the governance of agrifood systems (Schmitt 2020, 262). "A decentralization of power implies a construction of greater local autonomy in relation to vertical regulatory mechanisms imposed by the Food Empires," that is, the globalized agro-industrial system (Schmitt 2020, 271). In all those ways, EcoSol-agroecology initiatives devise "creative strategies to mobilise resources, capacities and connections in producing new forms of organizing labour." In particular, they develop capacities to generate employment and access finance, toward "a greater autonomy in the face of markets" (Schmitt 2010, 56, 60).

Policy Changes between 2016 and 2024

Since 2016, Brazil's right-wing federal governments have degraded or dismantled support measures for a solidaristic agroecology (Niederle et al. 2019; Sabourin, Craviotti, and Milhorance 2020). Budgets were reduced for measures run by the National Institute of Colonization and Agrarian Reform (INCRA), likewise for territorial development policies (PRONAT), which formerly had decentralized management by state and nonstate local actors (Sabourin, Craviotti, and Milhorance 2020). Some support measures were shifted to forms that exclude civil society and transfer control to agribusiness interests (Niederle, dos Santos, and Montiero 2021).

Resisting such changes, civil society countermovements reaffirmed an agenda for "political agroecology" as a form of political ecology (van der Ploeg 2021). They have campaigned for bottom-up support measures from municipal and state authorities. Within wider solidaristic networks for artisanal production, agroecology initiatives have promoted policy agendas and new statutes for municipal management of EcoSol policies (Rede de Gestores 2021). They have obtained such support in many localities, promoting horizontal *diálogos de saberes*, also called *ecologia de saberes* (de Sousa Santos 2007, 2009). In all those ways, civil society networks sought to revive or establish decentralized modes, initially at regional and state levels.

After the 2020 COVID-19 pandemic began, hygiene restrictions and labor shortages disrupted supply chains worldwide. Policy elites sought to restore, stabilize, and extend global commodity chains of the agro-industrial system (Clapp and Moseley 2020). Nevertheless, in many places mutual aid networks were extended or initiated to fill gaps in food provision, especially for vulnerable individuals. Although much was surplus from dominant food chains, some came from agroecological production. From a global perspective, "embedded, inclusive, often informal and unruly, economies, rooted in mutualism and solidarity, have flourished. . . . Food provisioning has seen a remarkable upsurge of solidarity and grassroots activism" (Leach et al. 2021, 7).

Did those solidaristic practices simply fill a temporary gap in the dominant system? Or did they also strengthen the basis for a broader long-term alternative? When the COVID-19 pandemic disrupted food supply chains and raised fears about virus contamination, many Brazilian consumers became more interested in food safety and quality. A Brazilian state agency provided information on local sources of healthy food, that is, produced with no agrochemicals and minimal processing (IDEC 2020).

EcoSol-agroecology networks made creative adaptations to maintain or even expand close relationships with consumers. Alongside virus protection and disinfection kits, means have included drive-through markets, alternative pickup points, box schemes, home deliveries, and so on. Predating the pandemic, community-supported agriculture was expanded through more subscriptions and new schemes, linking urban consumers with peri-urban or distant agroecological producers. These adaptations extended collective capacities that had previously been developed for *circuitos curtos*, as is well documented in Brazil (Calgaro, da Silva, and Santos 2022; Levidow, Sansolo, and Schiavinatto 2022). Such adaptations there have analogies in many countries of Latin America (see, e.g., Craviotti, Viteri, and Quinteros 2021; RIMISP 2020).

During Brazil's COVID-19 pandemic, *comida de verdade* (real food) became a more prominent slogan, signaling that pesticide-free agroecological methods were supporting people's health (ABA 2020). Hygiene requirements led to widespread school closures, jeopardizing agroecological purchases under the PNAE. So there were efforts to redirect the subsidy toward home deliveries, especially for vulnerable families, thus also maintaining producers' livelihoods (de Amorim, Ribeiro, and Bandoni 2020).

Collective marketing of agroecological products has been led mainly by women's associations (at least in Latin America). During the pandemic many such initiatives adapted to the hygiene requirements rather than suspending operations. They expanded virtual means for organizing and publicizing their activities. Members renewed their prepandemic bonds that had been built by talking, singing, reading poems, dancing, praying, embracing each other, and so on. This experience and memory facilitated agile adaptations by women's initiatives, strengthening their social resilience and inclusive role. This process occurred mainly at a local, informal, interpersonal level, thus remaining little known, according to a study by a feminist network. Likewise, thanks to interpersonal trust and familiarity, members agreed to be interviewed online for the study (SOF 2021, 31, 76).

As this illustrates, *sentipensante* capacities enabled agroecological initiatives to continue and likewise enabled activists to research the process. Such decentralized initiatives were linked with each other through solidaristic knowledge exchange, especially local or regional solidarity networks encompassing diverse artisanal activities. Through all these means, they built a collective capacity to design and implement adaptations during the pandemic.

Circuitos curtos have been resignifying traditional rural foods, especially in urban contexts during the pandemic. Agroecological products revive popular memory of agrifood traditions, thus restoring links between food knowledges and flavors, alliteratively called *saberes e sabores*. Going beyond economic relations, these efforts build affective sociocultural meanings that connect people from diverse backgrounds and localities (Menezes and Almeida 2021, 8, 92). The *feira do agricultor* has become "an opportune moment to amplify the debate over access to healthy food, and the educative role stimulated by proximity between farmer and consumer" (Menezes and Almeida 2021, 20).

Responding to the COVID-19 difficulties, all these efforts have used the opportunity for solidaristic aims. They strengthened the basis for a different future, rather than a return to normal. This ambition resonated with a continent-wide slogan against the neoliberal order: "No return to normality because normality was the problem!" (Fonseca 2020). After the 2022 general election brought back Luiz Inácio Lula da Silva to the presidency, his new government made several institutional improvements. It reestablished a Ministry for Agrarian Development and Family Agriculture (MDA) with new programs. The larger Agriculture Ministry added support measures for nonchemical production methods (MAPA 2022).

Moreover, it mainstreamed support measures for *circuitos curtos*, accommodating some demands from social movements and NGOs. In particular, it expanded the Programa Brasil Mais Cooperativo, which already was promoting cooperatives for collective marketing, especially for school meals. Its instruments encompass cooperation agreements and their decentralized implementation.

The government likewise revived the subsidy programs and agri-extension service (MAPA 2023b). The Family Agriculture Secretariat initiated a section for Cooperativismo e Soberania Alimentar (food sovereignty), in turn establishing the Mais Gestão program to strengthen family farmers' capacities for accessing markets (MDA 2023). However, *cooperativismo* encompasses a long-lasting conflict between capitalist enterprises versus solidaristic ones (see, e.g., Pelegrini, Shiki, and Shiki 2015).

The government extended a previous program, Bioeconomia Brasil Sociobiodiversidade, to promote sustainable development, people's well-being, and income generation. It has supported traditional communities and their small-scale enterprises sustainably using natural resources. It aims to conserve sociobiodiversity while using this basis for product supply chains, including herbal products, traditional foods, and Amazonian biomass (MAPA 2023a; ANA 2019a).

Such programs have offered new opportunities for an agroecological transformation, but alongside tensions with the hegemonic regime. The bioeconomy concept has featured long-time tension between the dominant capital-intensive form versus agroecological forms (Levidow, Birch, and Papaioannou 2013; Schmid, Padel, and Levidow 2012; TNI 2015). According to some Brazilian critics, agribusiness seeks a new extractivist frontier in forests, while marginalizing the Indigenous people who have cared for them (see, e.g., Coelho, 2023). Along those lines, the Lula government's agenda was criticized for bioeconomy resource appropriation lacking any relevance to agroecology. To shape public policies for agroecology, it will be necessary to expand popular participation, as well as to "conscientize the stomach, mind and heart" (ANA 2023).

For such alternatives, opportunities have come from two new programs, aiming especially to eliminate Brazil's widespread hunger and malnutrition, while also raising incomes of small-scale farmers. Plan Safra for Family Agriculture has aimed to increase production of healthy, pesticide-free food. Fresh or minimally processed foods would supply a basic food basket for distribution especially to vulnerable and needy families. The Solidarity Kitchens program has aimed to prepare and serve such food in cooperation with civil society groups.

The government has sought to expand agroecology by relaunching PLANAPO (CIAPO 2013). But this step was delayed by disputes over constraining agrochemicals. Within and beyond the government, agribusiness interests were defending intensive pesticide usage, including agrichemicals banned by the EU. Critics have diagnosed this threat as "chemical colonialism," highlighting Brazil's subaltern role in global export markets, regulatory regimes, and expert knowledge (ANA 2024; Bombardi 2023).

Conclusion

As the broader context, Latin America has had a sharp conflict between rival agendas. On one side, the hegemonic regime imposes a neocolonial plunder of people and nature by reducing them to labor and natural resources. Seeking global markets, this agenda is based on a centralized model of expert knowledge. On the other side, emerging from resistance, solidaristic alternatives promote a decentralized knowledge for democratizing the economy.

The latter effort has emerged jointly from social movements, practitioners, and academics, who together generated a decolonial framework known as Latin American critical thought. Some contributors have organized around networks for a solidarity economy or *economia solidaria* (EcoSol). Decentralized knowledge relates to solidaristic economic forms (de Sousa Santos and Rodriguez 2002). Socioeconomic decentralization needs workers' participation in their own decisions about production and distribution of its results, based on democratic self-governance (Monje-Reyes 2011, 720–21). EcoSol has had special prominence in Brazil, where such networks emerged from 1980s antijunta struggles, seeking both to democratize the economy and to decentralize knowledge.

Eventually, solidarity economy (EcoSol) movements converged with those promoting agroecological alternatives to the hegemonic agro-industrial system. These EcoSol-agroecology networks have decentralized and democratized knowledge by combining several sources, such as rural agrifood heritages, indigenous cultures, women's leadership, and mutual-aid traditions. Agroecosystems facilitate biodiverse resources, nutrient recycling, aesthetic qualities, and nutritional value.

By contrast with supermarket chains, *circuitos curtos* (short supply chains) have spread consumer knowledge about agroecological alternatives, while strengthening support for agroecological production methods. Pursuing these aims, initiatives build more secure livelihoods autonomous from exploitation and middlemen, while developing knowledge-exchange networks among groups devising such alternatives. In so doing, they contest and potentially overcome their subaltern role. This agenda has become especially prominent in Latin America, incorporating indigenous decolonial perspectives.

Agroecological innovation elaborates socioenvironmental technologies that are cheaply reproducible and easily adaptable to diverse contexts. Farmer–civil society partnerships draw on such local experiences for agroecological knowledge that can be generalized across diverse contexts. This convergence more effectively resists the dominant techno-diffusionist agri-modernization model.

In Brazil, the EcoSol-agroecology agenda demanded and gained state support measures for collective capacity building, especially during the left-wing governments of 2003–16. This agenda has strengthened *circuitos curtos*, which build consumer support for agroecological production methods and their results in artisanal products. Stimulated by feminist networks, women in particular found opportunities to gain or improve livelihoods through solidaristic cooperation. These efforts have sought to overcome inequalities of social class, gender, and race/ethnicity.

Between 2016 and 2020, Brazil's right-wing governments reduced or abolished those state support measures. Nevertheless, EcoSol-agroecology initiatives have continued through their prior collective capacities and knowledge-exchange processes (*diálogo de saberes*). Such capacities were extended to new initiatives through solidarity networks and some local authorities. They have sought to extend or revive horizontal decentralized knowledge networks that had already built collective capacities. Moreover, these strengthened demands for support measures on a territorial-regional basis, a bottom-up means to keep such measures accountable to solidarity networks. In this way, the network could build a regionalism from below, complementing efforts elsewhere in Latin America (Rosset et al. 2021). This strategy provided a basis to demand horizontal support measures in the new Lula administration from 2022 onward. Yet such efforts encountered many obstacles and internal conflicts (see previous section).

In all those ways, EcoSol networks have made a dual effort to democratize the economy and decentralize knowledge through epistemic alternatives. This linkage recasts the global dimension as social movements exchanging knowledges and strategies for counterhegemonic, decolonial aims. This brings a potential for a postcapitalist socioeconomic transformation, contrary to the hegemonic capitalist transformation through the centralized techno-diffusionist model.

Outcomes will depend on many struggles such as: civil society networks shaping and decentralizing EcoSol-agroecology support measures; state agencies gaining greater political-technical capacities for such measures; popular resistance against the hegemonic agrifood regime alongside its neocolonial extractivist basis; and a contest over this regime's capital-intensive epistemic centralization.

Note

This chapter comes from the AgroEcos project Research Partnership for an Agroecology-Based Solidarity Economy in Bolivia and Brazil, AH/ T004274/1, which ran during 2020–22, funded by the United Kingdom's Arts and Humanities Research Council (AHRC), Global Challenges Research Fund (GCRF), https://projetoagroecos.wixsite.com/meusite. Thanks to all the project's teams for insights informing this chapter, especially the UNESP core team of Davis Sansolo and Monica Schiavinatto. Thanks also to this book's editors for their helpful suggestions. Likewise to Paulo Andre Niederle as regards the post-2022 Lula presidency.

References

ABA. 2020. "Comida de Verdade: Agricultura familiar segue pro-
duzindo saúde em época de pandemia." Associação Brasileira
de Agroecologia (ABA), posted to Facebook, March 30. https://
www.facebook.com/articulacaonacionaldeagroecologia/videos
/agricultura-familiar-produ%C3%A7%C3%A3o-de-alimentos-e
-comercializa%C3%A7%C3%A3o-em-%C3%A9poca-de-pandemi
/215572959762195/.

ActionAid Brasil. 2010. *Mulheres e agroecologia: Sistematização de experiências de mulheres agricultoras*. Rio de Janeiro: Articulação Nacional de Agroecologia. http://actionaid.org.br/wp-content/files_mf
/1584562583MulhereseAgroecologia.pdf.

AgroEcos. 2022. *EcoSol-Agroecology as Resistance and Alternative Development: Insights from South American Experience*. https://3d33eb12-f421
-47a1-a45f-76acc45bd2d6.filesusr.com/ugd/5872ec_a36390e0a69b4ac4b6
cfc48ee50645c6.pdf.

Alcoff, Linda Martín. 2022. "Extractivist Epistemologies." *Tapuya: Latin
American Science, Technology and Society* 5 (1): 212–31. https://doi.org/10
.1080/25729861.2022.2127231.

Altieri, Miguel A. 2002. *Agroecologia: Bases científicas para uma agricultura
sustentável*. Guaíba: Agropecuária.

Altieri, Miguel A., and Clara I. Nicholls. 2008. "Scaling Up Agroecological
Approaches for Food Sovereignty in Latin America." *Sustainable Agriculture Reviews* 11: 1–29.

Altieri, Miguel A., and Victor Manuel Toledo. 2011. "The Agroecological
Revolution in Latin America: Rescuing Nature, Ensuring Food Sovereignty and Empowering Peasants." *Journal of Peasant Studies* 38 (3):
587–612.

ANA. 2007. *Construção do conhecimento agroecológico: Novos papéis,
novas identidades*. Caderno do II Encontro Nacional de Agroecologia (June 2007). Guarapari: Articulação Nacional de Agroecologia
(ANA). https://aspta.org.br/files/2020/04/Construcao-Conhecimento
-Agroecol%C3%B3gico-Novos-Pap%C3%A9is-Novas-Identidades-ANA
-2007.pdf.

ANA. 2012. *Propostas da Articulação Nacional de Agroecologia-ANA para a
"Politica Nacional de Agroecologia e Produção Organica."* Rio de Janeiro:
Articulação Nacional de Agroecologia (ANA). https://agroecologia
.org.br/wp-content/uploads/2012/12/propostas-da-ana-para-a-politica
-nacional-de-agroecologia-e-producao-organica-pnapo.pdf.

ANA. 2019a. *Agro-socio biodiversidade: Direitos, democracia e agroecologia no
campo e na ciudade*. Rio de Janeiro: Articulação Nacional de Agroecologia
(ANA). https://www.biodiversidadla.org/Documentos/Agro-socio
-biodiversidade-Direitos-democracia-e-agroecologia-no-campo-e-na
-ciudade.

ANA. 2019b. *Redes de agroecologia para o desenvolvimento dos territórios: Aprendizados do Programa Ecoforte; Sumário Executivo.* Edited by Anna Cecília Cortines. Rio de Janeiro: Articulação Nacional de Agroecologia (ANA).

ANA. 2023. "A agroecologia é o principal caminho para o desenvolvimento agrário." *Seminário Políticas Públicas de Agroecologia na Boca do Povo*, November 22. https://agroecologia.org.br/2023/11/22/a-agroecologia-e-o-principal -caminho-para-o-desenvolvimento-agrario-diz-fernanda-machiaveli-do-mda -no-seminario-politicas-publicas-de-agroecologia-na-boca-do-povo/.

ANA. 2024. "'Colonialismo químico pode explicar dificuldades na redução de agrotóxicos,' diz Larissa Bombardi." Articulação Nacional de Agroecologia (ANA). https://agroecologia.org.br/2024/08/13/colonialismo -quimico-pode-explicar-dificuldades-na-reducao-de-agrotoxicos-diz -larissa-bombardi/.

Barbosa, Lia Pinheiro. 2016. "Educación, resistencia y conocimiento en América Latina: Por una teoría desde los movimientos sociales." *Revista de Raíz Diversa* 3 (6): 45–79. https://doi.org/10.22201/ppela.24487988e .2016.6.58425.

Barbosa, Lia Pinheiro. 2019. "Estética da resistência: Arte sentipensante e educação na práxis política dos movimentos camponeses e indígenas latino-americanos." Dossiê Educação, Arte e Política: experiências, alternativas e resistências. *Revista Conhecer* 10 (23): 29–62.

Bolivia. 2009. Constitución Política del Estado del Estado, Febrero de 2009. Servicio Estatal de Autonomías. https://sea.gob.bo/digesto /CompendioNormativo/01.pdf.

Bombardi, Larissa Mies. 2023. *Agrotóxicos e colonialismo químico.* São Paulo: Elefante.

Borsatto, Ricardo Serra, Sonia Maria Pessoa Pereira Bergamasco, and Valter Bianchini. 2017. *Transferência de tecnologia ou compartilhamento de conhecimentos? Desvendando o papel da Embrapa no desenvolvimento rural.* Brasília: Empresa Brasileira de Pesquisa Agropecuária (Embrapa).

Brazil. 1988. Constituição da República Federativa do Brasil. Congresso Nacional, normas.leg.br. https://www.senado.leg.br/atividade/const /con1988/con1988_05.10.1988/art_204_.asp.

Brazil. 2009. Lei No. 11.947, Programa Nacional de Alimentação Escolar (PNAE). June 16.

Cabral, Lídia, Arilson Favareto, Langton Mukwereza, and Kojo Amanor. 2016. "Brazil's Agricultural Politics in Africa: More Food International and the Disputed Meanings of 'Family Farming.'" *World Development* 81: 47–60.

Calgaro, Hemerson Fernandes, Newton José Rodrigues da Silva, and Wagner Santos. 2022. *Circuitos alimentares de proximidades e a economia solidária.* Documento Técnico 130. Campinas: Governo do Estado de São Paulo, Secretaria de Agricultura e Abastecimento, Coordenadoria de Assistência Técnica Integral—CATI. https://www.cati.sp.gov.br/portal /themes/unify/arquivos/produtos-e-servicos/acervo-tecnico/CIRCUIT OSALIMENTARESECONOMIASOLIDARIA%206_12_2022_c.pdf.

Caporal, Francisco Roberto, and José Antônio Costabeber. 2004. *Agroecologia e extensão rural: Contribuições para a promoção do desenvolvimento rural sustentável*. Brasília: EMATER.

CIAPO. 2013. *Brasil agroecologico: Plano Nacional de Agroecologia e Produção Orgânica—PLANAPO*. Brasília: Ministério do Desenvolvimento Agrário, Câmara Interministerial de Agroecologia e Produção Orgânica (CIAPO) and Proyecto Alianza por la agroecología. https://agroecologia.org.br /wp-content/uploads/2013/11/planapo-nacional-de-agroecologia-e -producao-organica-planapo.pdf.

Clapp, Jennifer, and William G. Moseley. 2020. "This Food Crisis Is Different: COVID-19 and the Fragility of the Neoliberal Food Security Order." *Journal of Peasant Studies* 47 (7): 1393–1417.

CNES. 2015. *1º Plano Nacional de Economia Solidária (2015–2019): Para promover o direito de produzir e viver de forma associativa e sustentável*. Brasília: Conselho Nacional de Economia Solidária (CNES).

Coelho, Luisa. 2023. "'Bioeconomia,' novo flanco de exploração do agronegocio." *OutrasMídias*, November 8. https://outraspalavras.net/outrasmidias /bioeconomia-novo-flanco-de-exploracao-do-agronegocio/.

Coraggio, Jose Luis. 2016. "La economía social y solidaria (ESS): Niveles y alcances de acción de sus actors: El papel de las universidades." In *Economía social y solidaria: Conceptos, prácticas y políticas públicas*, edited by Carlos Puig, 15–40. Valparaíso: Universidad de Valparaíso.

Craviotti, Clara, María Laura Viteri, and Gladys Quinteros. 2021. "Covid-19 y circuitos cortos de comercialización de alimentos en Argentina: El papel de los actores sociales." *European Review of Latin American and Caribbean Studies*, no. 112: 29–49. https://doi.org/10.32992/erlacs.10780.

Dagnino, Renato. 2010. *Tecnologias sociais: Ferramenta para construir outra sociedade*. 2nd ed. Campinas: Unicamp. https://cdt.unb.br/images /CEDES/2010_FERRAMENTA_TEC_SOCIAL_LIVRO.pdf.

Darolt, Moacir R., Claire Lamine, and Alfio Brandemburg. 2013. "A diversidade dos circuitos curtos de alimentos ecológicos: Ensinamentos do caso brasileiro e francês." *Agriculturas* 10 (2): 8–13.

Dash, Anup. 2014. *Toward an Epistemological Foundation for Social and Solidarity Economy*. Geneva: United Nations Research Institute for Social Development (UNRISD).

Dash, Anup. 2016. "An Epistemological Reflection on Social and Solidarity Economy." *Forum for Social Economics* 45 (1): 61–87.

de Amorim, Ana Laura Benevenuto, José Raimundo Sousa Ribeiro Jr., and Daniel Henrique Bandoni. 2020. "Programa Nacional de Alimentação Escolar: Estratégias para enfrentar a insegurança alimentar durante e após a COVID-19." *Revista de Administração Pública* 54 (4): 1134–45.

Delgado, Freddy, and Stephane Rist. 2016. *Ciências, diálogo de saberes y transdisciplinariedad: Aportes teórico metodológicos para sustentabilidade alimentaria y del desarollo*. La Paz: Agruco.

de Sousa Santos, Boaventura. 2007. "Para além do pensamento abissal: Das linhas globais a uma ecologia de saberes." *Novos Estudos—CEBRAP* 79: 71–94. https://doi.org/10.1590/S0101-33002007000300004.

de Sousa Santos, Boaventura. 2009. *Pensar el estado y la sociedad: Desafíos actuales.* Buenos Aires: Waldhuter Editores.

de Sousa Santos, Boaventura, and C. Rodriguez. 2002. Introdução: Para ampliar o cânone da produção. Translated by Vítor Ferreira. In *Produzir para viver: Os caminhos da produção não capitalista,* edited by B. de Sousa Santos, 23–77. Rio de Janeiro: Civilização Brasileira.

dos Santos, Aline Mendonça, and Vanderson Gonçalves Carneiro. 2008. "O movimento da economia solidária no Brasil: Uma discussão sobre a possibilidade da unidade através da diversidade." *e-cadernos CES* 2. https://journals.openedition.org/eces/1260.

Ecovida. 2003. "Participação da rede no marco regulatório da lei federal." Três Cachoeiras-RS: Rede de Agroecologia Ecovida. https://www.ecovida.org.br/nossas_acoes/certificacao/lei_rede/.

Fals-Borda, Orlando. 2008. *Una sociología sentipensante para América Latina.* Buenos Aires: CLACSO/Siglo del Hombre Editores.

FBES. 2011. *Encontro Nacional de Diálogos e Convergências: Agroecologia, saúde e justiça ambiental, soberania alimentar, economia solidária e feminism.* Brasília: Forum Brasileiro de Economia Solidária (FBES). https://fbes.org.br/wp-content/uploads/Acervo/Publica%C3%A7%C3%B5es/fbes_encontro_dialogo_e_convergencias_documento_referencia.pdf.

FBES. 2012. "Comissão Organizadora da V Plenária Nacional de Economia Solidária." Brasília: Fórum Brasileiro de Economia Solidária (FBES).

Fonseca, Iara. 2020. "Não podemos voltar ao normal, porque o normal era o problema!" *Resiliência,* March 31. https://www.resilienciamag.com/nao-podemos-voltar-ao-normal-porque-o-normal-era-o-problema/.

Fressoli, M., and R. Dias. 2014. *The Social Technology Network: A Hybrid Experiment in Grassroots Innovation.* STEPS Working Paper 67. Brighton: STEPS Centre. http://steps-centre.org/wp-content/uploads/Social-Technology1.pdf.

Grisa, Catia, Claudia Job Schmitt, Lauro Francisco Mattei, Renato Sergio Maluf, and Sergio Pereira Leite. 2009. "O Programa de Aquisição de Alimentos (PAA) em perspectiva: Apontamentos e questões para o debate." *Retratos de Assentamentos* 13 (1): 137–70.

Grosfoguel, Ramón. 2019. "Epistemic Extractivism: A Dialogue with Alberto Acosta, Leanne Betasamosake Simpson, and Silvia Rivera Cusicanqui." In *Knowledges Born in the Struggle,* edited by Boaventura de Sousa Santos and Maria Meneses, 203–18. London: Routledge.

Guanziroli, Carlos, Antonio Buainain, and Alberto Sabbato. 2013. "Family Farming in Brazil: Evolution between the 1996 and 2006 Agricultural Censuses." *Journal of Peasant Studies* 40 (5): 817–43.

GVC. 2015. *Plantas Alimentícias Não Convencionais (PANCS): Hortaliças espontâneas e nativas*. Porto Alegre: Grupo Viveiros Comunitários (GVC), Universidade Federal do Rio Grande do Sul.

Haesbaert, Rogério. 2021. *Território e descolonialidade: Sobre o giro (multi)territorial/de(s)colonial na América Latina*. Buenos Aires: CLACSO, Programa de Pós-Graduação em Geografia, Universidade Federal Fluminense.

Hillenkamp, Isabelle, Isabelle Guérin, and Christine Verschuur. 2014. "Economie solidaire et theories feministes: Pistes pour une convergence necessaire." *Revista de Economia Solidaria*, ACEESA: 4–43, ird-01197164. https://base.socioeco.org/docs/hillenkamp_20140130_texte_final.pdf.

Hillenkamp, Isabelle, Isabelle Guérin, and Christine Verschuur. 2017. "Cruzando os caminhos da economia solidária e do feminismo: Passos para uma convergência necessária." *Revista economía* 69 (109): 43–59.

HLPE. 2013. *Investing in Smallholder Agriculture for Food Security*. Rome: FAO, High-Level Panel of Experts on Food Security and Nutrition of the Committee on World Food Security (HLPE). https://www.fao.org/4/i2953e/i2953e.pdf.

HLPE. 2019. *Agroecological and Other Innovative Approaches for Sustainable Agriculture and Food Systems That Enhance Food Security and Nutrition*. HLPE Report 14. Rome: FAO, High-Level Panel of Experts on Food Security and Nutrition of the Committee on World Food Security (HLPE). https://openknowledge.fao.org/server/api/core/bitstreams/ff385e60 -0693-40fe-9a6b-79bbef05202c/content.

HLPE. 2020. *Food Security and Nutrition: Building a Global Narrative towards 2030*. HLPE Report 15. Rome: FAO, High Level Panel of Experts on Food Security and Nutrition of the Committee on World Food Security (HLPE). https://openknowledge.fao.org/server/api/core/bitstreams /8357b6eb-8010-4254-814a-1493faaf4a93/content.

Holt-Giménez, Eric. 2006. *Campesino a Campesino: Voices from Latin America's Farmer to Farmer Movement for Sustainable Agriculture*. Oakland, CA: Food First.

IAM. 2016. *Laboratório de experimentação em tecnologias Socioambientais (LETS)*. Rio de Janeiro: Instituto Ambiente em Movimento (IAM).

IBGE. 2006. *Censo Agropecuário 2006: Agricultura familiar; primeiros resultados*. Brasília: Instituto Brasileiro de Geografia e Estatstica (IBGE).

IBGE. 2017. *Censo Agropecuário 2017: Agricultura familiar; primeiros resultados*. Brasília: Instituto Brasileiro de Geografia e Estatstica (IBGE).

IDEC. 2020. "Idec cria plataforma para divulgar iniciativas que comercializam alimentos saudáveis durante a pandemia do Covid-19, 6 abril." Instituto de Defesa de Consumidores (IDEC), April 14. https://idec.org.br/release /idec-cria-plataforma-para-divulgar-iniciativas-que-comercializam -alimentos-saudaveis-durante.

ITS. 2004. *Caderno de debate: Tecnologia social no Brasil*. São Paulo: Instituto de Tecnologia Social (ITS). https://drive.google.com/file/d /1pSB4DlxqDIiBL15M3epNToby-yDnQ5Xo/view.

Kay, Sylvia, ed. 2016. *Connecting Smallholders to Markets: An Analytical Guide*. Rome: Civil Society Mechanism. https://www.csm4cfs.org/wp -content/uploads/2016/10/ENG-ConnectingSmallholdersToMarkets _web.pdf.

Lander, Edgardo. 2001. "Pensamiento crítico latinoamericano: La impugn-ación del eurocentrismo." *Revista de Sociología* 15: 13–25. https://doi.org /10.5354/0719-529X.2001.27766.

Leach, Melissa, Hayley MacGregor, Ian Scoones, and Annie Wilkinson. 2021. "Post-pandemic Transformations: How and Why COVID-19 Requires Us to Rethink Development." *World Development* 138: 105233.

Leff, Enrique. 2001. *Saber ambiental: Sustentabilidade, racionalidade, com-plexidade, poder*. Petrópolis, Rio de Janeiro: Vozes.

Leff, Enrique. 2009. "Degrowth, or Deconstruction of the Economy: Towards a Sustainable World." Occasional Paper Series No. 6. *Critical Currents* 6: 101–7. Uppsala: Dag Hammarskjöld Foundation.

Levidow, Les, Kean Birch, and Theo Papaioannou. 2013. "Divergent Paradigms of European Agro-Food Innovation: The Knowledge-Based Bio-economy (KBBE) as an R&D Agenda." *Science, Technology and Human Values* 38 (1): 94–125. http://journals.sagepub.com/doi/abs/10.1177 /0162243912438143.

Levidow, Les, Davis Sansolo, and Monica Schiavinatto. 2022. "EcoSol-Agroecology Networks Respond to the Covid-19 Crisis in Brazil's Baixada Santista Region." *Journal of Peasant Studies* 49 (7): 1409–45. https://www .tandfonline.com/doi/full/10.1080/03066150.2022.2096447.

Lobo, Natália. 2021. *As novas formas dos velhos mecanismos de apropriação da natureza: Controle dos corpos-tempos-territórios e política feminista*. São Paulo: SOF Sempreviva Organização Feminista. https://www.sof.org .br/wp-content/uploads/2021/08/Novas_formas_velhos_mecanismos -Natalia_Lobo-SOF.pdf.

Long, Norman, and Jan Douwe van der Ploeg. 2011. "Heterogeneidade, ator e estrutura: Para a reconstituição do conceito de estrutura." In *Os atores do desenvolvimento rural: Perspectivas teóricas e práticas sociais*, ed-ited by Sergio Schneider and Marcio Gazilla, 21–48. Porto Alegre: PGDR/ UFRGS.

Magnanti, Natal João. 2008. "Circuito Sul de circulação de alimentos da Rede Ecovida de Agroecologia." *Agriculturas* 5 (2): 26–29.

MAPA. 2007. "Decreto Nº 6.323, de 27 de dezembro de 2007. Regulamenta a Lei no 10.831, de 23 de dezembro de 2003, que dispõe sobre a agricultura orgânica, e dá outras providências." Brasília: Ministerio de Agricultura, Pecuária e Abastecimento (MAPA). https://www.gov.br/agricultura/pt -br/assuntos/sustentabilidade/organicos/legislacao/portugues/decreto -no-06-323-de-27-de-dezembro-de-2007.pdf/view.

MAPA. 2008. *Controle social na venda direta ao consumidor de produtos orgânicos sem certificação*. Brasília: Ministério da Agricultura, Pecuária e Abastecimento (MAPA).

MAPA. 2022. "Tecnologias de cultivo de orgânicos está disponível no site do Mapa." Ministerio de Agricultura, Pecuária e Abastecimento (MAPA), November 10. https://www.gov.br/agricultura/pt-br/assuntos/noticias /tecnologias-de-cultivo-de-organicos-esta-disponivel-no-site-do-mapa.

MAPA. 2023a. *Bioeconomia Brasil—Sociobiodiversidade.* Brasília: Ministério da Agricultura, Pecuária e Abastecimento (MAPA). https://www.gov.br /agricultura/pt-br/assuntos/camaras-setoriais-tematicas/documentos /camaras-setoriais/hortalicas/2019/58a-ro/bioeconomia-dep-saf-mapa.pdf.

MAPA. 2023b. *Brasil Mais Cooperativo.* Brasília: Ministério da Agricultura, Pecuária e Abastecimento(MAPA). https://ceades.org.br/wp-content /uploads/2023/07/Programa-Brasil-Mais-Cooperativo-Anater.pdf

Matte, Alessandra, and Potira Viegas Preiss. 2019. "Protagonismo de produtores e consumidores na construção de mercados alimentares sustentáveis." In *Alimentação e sustentabilidade,* edited by Rodrigo Machado Vilani, Elidio Vanzella, and Adriana Bambrilla, 125–56. João Pessoa, Brazil: Centro de Comunicação, Turismo e Artes (CCTA), Universidade Federal da Paraíba.

MDA. 2023. *MDA lança programa de gestão para cooperatias da agricultura familiar.* Brasília: Ministério do Desenvolvimento Agrário e Agricultura Familiar (MDA). https://agenciabrasil.ebc.com.br/economia/noticia/2023-08/mda -lanca-programa-de-gestao-para-cooperativas-da-agricultura-familiar.

Medina, Gabriel, Camila Almeida, Evandro Novaes, Javier Godar, and Benno Pokorny. 2015. "Development Conditions for Family Farming: Lessons from Brazil." *World Development* 74 (C): 386–96.

Menezes, Sônia de Souza Mendonça, and Maria Geralda de Almeida, eds. 2021. *Vamos às feiras! Cultura e ressignificação dos circuitos curtos.* Aracaju: Criação Editora. https://editoracriacao.com.br/wp-content/uploads /2021/08/feirassite.pdf.

Mignolo, Walter. 2000. *Local Histories/Global Designs: Coloniality, Subaltern Knowledges, and Border Thinking.* Princeton, NJ: Princeton University Press.

Mignolo, Walter, and Catherine E. Walsh. 2008. *On Decoloniality: Concepts, Analytics, Praxis.* Durham, NC: Duke University Press.

Monje-Reyes, Pablo. 2011. "Economía solidaria, cooperativismo y descentralización: La gestión social puesta en práctica." *Cadernos EBAPE.BR* 9 (3): 704–23. https://bibliotecadigital.fgv.br/ojs/index.php/cadernosebape /article/view/5216/3950.

Mussoi, E. M. 2011. *Política de Extensión Rural Agroecológica en Brasil: Avances y desafíos en la transición en las instituciones oficiales.* Cordoba: Universidad de Cordoba. https://wp.ufpel.edu.br/consagro/files/2011 /12/MUSSOI-Eros-Marion-Pol%C3%ADtica-de-Extensi%C3%B3n -Rural-Agroecol%C3%B3gica-en-Brasil-avances-y-desaf%C3%ADos-en-la -transici%C3%B3n-en-las-instituciones-oficiales.pdf.

Niederle, Paulo André, Catia Grisa, Everton Lazaretti Picolotto, and Denis Soldera. 2019. "Narrative Disputes over Family-Farming Public Policies in Brazil: Conservative Attacks and Restricted Countermovements." *Latin American Research Review* 54 (3): 707–20. http://doi.org/10.25222/larr.366.

Niederle, Paulo André, Rodrigo Salles Pereira dos Santos, and Cristiano Fonseca Monteiro. 2021. "Interpretações institucionalistas sobre as transformações dos capitalismos brasileiros: Da pretensão neodesenvolvimentista à predação." *Revista Brasileira de Sociologia* 9 (22): 9–44.

Nobre, Miriam, Nalu Faria, and Renata Moreno. 2015. *Las mujeres en la construcción de la economía solidaria y la agroecología: Textos para la acción feminista.* São Paolo: Sempreviva Organização Feminista (SOF). https://www.socioeco.org/bdf_fiche-document-4421_en.html.

Pelegrini, Djalma Ferreira, Simone de Faria Narciso Shiki, and Shigeo Shiki. 2015. "Uma abordagem teórica sobre cooperativismo e associativismo no Brasil." *Revista Eletronica Extensao* 12 (19): 70–85. http://dx.doi.org/10.5007/1807-0221.2015v12n19p70.

Petersen, Paulo, Eros Marion Mussoi, and Fabio Dal Soglio. 2013. "Institutionalization of the Agroecological Approach in Brazil: Advances and Challenges." *Agroecology and Sustainable Food Systems* 37 (1): 103–14.

PIADAL. 2013. *Agricultura y desarrollo en América Latina: Gobernanza y políticas públicas.* Buenos Aires: TESEO, Panel Independiente sobre Agricultura para el Desarrollo de América Latina (PIADAL).

Pires, J. H. S., and H. T. Novaes. 2016. "Estudo, trabalho e agroecologia: A proposta política pedagógica dos Cursos de Agroecologia do mst no Paraná." *Germinal: Marxismo e Educação em Debate* 8 (2): 110–24. https://doi.org/10.9771/gmed.v8i2.18144.

Rede de Gestores de Políticas Públicas de Economia Solidária. 2008. *Políticas públicas em economia solidária: Reflexões da rede de gestores.* Centro de Estudos e Pesquisa Josué de Castro; Secretaria Nacional de Economia Solidária. Recife: Ed. Universitária da UFPE.

Rede de Gestores de Políticas Públicas de Economia Solidária. 2021. "Oficina de Planejamento da Economia Solidária na gestão municipal." https://www.rededegestoresecosol.org.br/oficina-de-planejamento-economia-solidaria-nas-prefeituras/.

RIMISP. 2020. *Análisis de coyuntura COVID-19 en América Latina.* N.p.: Centro Latinoamericano para el Desarrollo Rural (RIMISP). https://www.rimisp.org/wp-content/uploads/2020/04/02-Covid-19-Abastecimiento.pdf.

RIPESS. 2012. "La economía que necesitamos: Declaración del movimiento de Economía Social y Solidaria a Rio +20." Réseau Intercontinental de Promotion de L'économie Sociale Solidaire (RIPESS), June. http://rio20.net/wp-content/uploads/2012/08/2012_declaration_ripess_rio_20_esp.pdf.

RIPESS. 2015. "Visión global de la economía social solidaria: Convergencias y contrastes en los conceptos, definiciones y marcos conceptuales." Réseau Intercontinental de Promotion de L'économie Sociale Solidaire (RIPESS), February. https://base.socioeco.org/docs/ripess_vision_global_esp.pdf.

Rosset, Peter M., Lia Pinheiro Barbosa, Valentín Val, and Nils McCune. 2021. "Critical Latin American Agroecology as a Regionalism from Below." *Globalizations* 19 (4): 635–52. https://doi.org/10.1080/14747731.2021.1923353.

Rosset, Peter M., and M. E. Martínez-Torres. 2012. "Rural Social Movements and Agroecology: Context, Theory, and Process." *Ecology and Society* 17 (3). https://doi.org/10.5751/ES-05000-170317.

Sabourin, Eric. 2011. *Sociedades e organizações camponesas: Uma leitura através da reciprocidade.* Porto Alegre: UFRGS/PGDR.

Sabourin, Eric, Clara Craviotti, and Carolina Milhorance. 2020. "The Dismantling of Family Farming Policies in Brazil and Argentina." *International Review of Public Policy* 2 (1): 45–67. https://journals.openedition.org/irpp/799.

Schmid, Otto, Susanna Padel, and Les Levidow. 2012. "The Bio-economy Concept and Knowledge Base in a Public Goods and Farmer Perspective." *Bio-based and Applied Economics (BAE)* 1 (1): 47–63. http://orgprints.org/20942/.

Schmitt, Claudia Job. 2010. "Economia solidária e agroecologia: Convergências e desafios na construção de modos de vida sustentáveis." *Mercado de trabalho* 42: 55–64. http://repositorio.ipea.gov.br/bitstream/11058/4050/1/bmt42_08_Eco_02_convergencias.pdf.

Schmitt, Claudia Job. 2018. *Agro-socio-biodiversidade: Direitos, democracia e agroecologia no campo e na cidade.* Rio de Janeiro: Articulação Nacional de Agroecologia. https://www.biodiversidadla.org/Documentos/Agro-socio-biodiversidade-Direitos-democracia-e-agroecologia-no-campo-e-na-ciudade.

Schmitt, Claudia Job. 2020. *Redes de agroecologia para o desenvolvimento dos territórios: Aprendizados do Programa Ecoforte.* Rio de Janeiro: Articulação Nacional de Agroecologia.

Schmitt, Claudia Job, Paulo Niederle, Mario Ávila, Eric Sabourin, Paulo Petersen, Luciano Silveira, William Assis, Juliano Palm, and Gabriel B. Fernandes. 2017. "La experiencia brasilena de construccion de políticas públicas a favor de la Agroecologia." In *Políticas públicas a favor de la agroecología em América Latina y El Caribe*, edited by Eric Sabourin, Maria Mercedes Patrouilleau, Jean François Le Coq, Luis Vasquez, and Paulo Andre Niederle, 73–122. Porto Alegre: Evangraf/Criação Humana, Red PP-AL/FAO. https://openknowledge.fao.org/server/api/core/bitstreams/75414f26-70b4-40cd-9cf6-f34cdab8bf4a/content.

Schüttz, Gabriela D'Ávila, and Luiz Inácio Gaiger. 2006. "O mister de reaprender os vínculos entre a economia e a vida social." In *Educação e sócio-economia solidária: Interação universidade–movimentos sociais*, edited by Laudemir Luiz Zart and Josivaldo Constantino dos Santos, 50–64. Cáceres-MT: Editora Unemat. http://www.ctamt.org.br/storage/publicacao/publicacao/educacao-e-socio-economia-solidaria-livro.pdf.

Schwab, Fabio, and Ángel Calle Collado. 2017. *Experiencias agroecológicas de comercialización en Brasil.* Paper no. 58, El Futuro de la Alimentación y Retos de la Agricultura para el Siglo XXI: elikadura. https://www.elikadura21.ehnebizkaia.eus/wp-content/uploads/2017/04/58-Schwab.pdf.

Serafim, M. P., V. M. B. de Jesus, and J. Faria. 2013. "Tecnologia social, agro-ecologia e agricultura familiar: Análises sobre um processo sociotécnico." *Segurança Alimentar e Nutricional, Campinas* 20, no. 1 (suppl.): 169–81. https://doi.org/10.20396/san.v20i1supl.8634595.

Singer, Paul, and André Ricardo Souza, eds. 2000. *A economía solidária no Brasil: A autogestão como resposta ao desemprego.* São Paulo: Contexto.

SOF. 2015. *Las mujeres en la construcción de la economía solidaria y la agro-ecología: Textos para la acción feminista.* São Paulo: Sempreviva Organiza-ção Feminista (SOF). https://www.sof.org.br/wp-content/uploads/2015/12/Livro-Agroecologia-web.pdf.

SOF. 2021. *Um meio tempo preparando outro tempo: Cuidados, produção de ali-mentos e organização de mulheres agroecológicas na pandemia.* São Paulo: Sempreviva Organização Feminista (SOF). https://www.sof.org.br/wp-content/uploads/2021/04/210407_ummeiotempo_sof_08_rev.pdf.

TNI. 2015. *The Bioeconomy: A Primer.* Amsterdam: Transnational Institute (TNI).

Tonneau, Jean-Philippe, Joacir Rufino de Aquino, and Olívio Alberto Teixeira. 2005. "Modernização da agricultura familiar e exclusão social: O dilema das políticas agrícolas." *Cadernos de Ciência e Tecnologia* 22 (1): 67–82.

Utting, Peter. 2018. *Achieving the Sustainable Development Goals through Social and Solidarity Economy: Incremental versus Transformative Change.* Geneva: United Nations Research Institute for Social Development.

van der Ploeg, Jan Douwe. 2021. "The Political Economy of Agroecology." *Journal of Peasant Studies* 48 (2): 274–97.

van der Ploeg, Jan Douwe, and Gert van Dijk, eds. 1995. *Beyond Moderniza-tion: The Impact of Endogenous Development.* Assen, Netherlands: Van Gorcum.

van der Ploeg, Jan Douwe, Ye Jingzhong, and Sergio Schneider. 2012. "Rural Development through the Construction of New, Nested, Markets: Comparative Perspectives from China, Brazil and the European Union." *Journal of Peasant Studies* 39 (1): 133–73.

Weber, Max. 1930. *The Protestant Ethic and the Spirit of Capitalism.* Trans-lated by Talcott Parsons. London: Routledge.

Zaoual, Hassan. 2010. "O homo situs e suas perspectivas paradigmáticas." *Oikos* 9 (1): 13–39.

Zuluaga Sánchez, Gloria Patricia, Georgina Catacora-Vargas, and Emma Siliprandi. 2018. *Agroecología en femenino: Reflexiones a partir de nuestras experiencias.* La Paz: SOCLA (CLACSO).

III Creating Alternative Spaces

Therapeutic Space as Knowledge Space

Decentralizing Biomedicine in Inpatient Hospice and Palliative Care

Sensing Invisible Knowledge Spaces in Taiwan's Palliative Care

Responding to the challenge that "decentering is, in short, a reconfiguration of dominant or hegemonic ideas in light of new ideas that challenge and correct them, giving rise to strong objective knowledge," and demonstrating how epistemic decentralizing reorients flows of thinking by "empowering peripheral actors, infrastructures and practices in such a way that they can produce knowledge of themselves," this chapter aims to achieve a better understanding of hospital-based hospice and palliative care in Taiwan, an East Asian state that has been praised as "offer[ing] a good place to die" (Duke-NUS Medical School 2022).

Let me briefly introduce the study that reached that finding. "Cross Country Comparison of Expert Assessments of the Quality of Death and Dying 2021" is the third in a series of comparative studies since 2010 on hospice care worldwide, sponsored by the Lien Foundation in Singapore (Duke-NUS Medical School 2022). In the previous two studies, the researchers used twenty quantitative and qualitative indicators across five categories of weighting—palliative and health care environment, human resources, affordability of care, quality of care, and community engagement—to compare each state's performance in offering quality palliative care.[1]

Originally scheduled for 2020 but postponed due to the COVID-19 pandemic, the research was conducted by a team at Duke–National University of Singapore (NUS) Medical School and the research method modified.[2] According to its principal investigator, Eric Finkelstein, the study design was improved by identifying seven domains and thirty-three subdomains that capture the key aspects of quality of death and by better weighting of these indicators in accord with a patient-centered approach. As a result, this study was able to garner expert opinions, from eighty-one countries, on the preferences and considerations that matter most to patients and families at end of life.

Taiwan's performance shone in these studies. In the first two, it was ranked fourteenth and sixth in the world respectively, and highest among all Asian states. When evaluated using the new methodology in this latest study, Taiwan reached third place, its best ranking ever, and was recognized as one of six states to enjoy an A-grade score.

There are institutional hints for interpreting Taiwan's success. As indicated in the Duke-NUS study, high-income states tend to provide a better quality of palliative care, while their low-income counterparts, suffering under the high cost of such care, do comparatively better in nonmedical concerns and spiritual needs. Taiwan's performance falls into neither of these two categories. Listed as a high-income economy and with a high-quality medical workforce, Taiwan's health authority commits to palliative care and includes this in its national health insurance scheme, making hospice a quality and affordable care option. What is equally important is the establishment of laws and policies concerning hospices. Following discussions among medical professionals in the 1990s, in 2000 the Hospice Palliative Care Regulations (安寧緩和醫療條例) were promulgated, making Taiwan the first state in Asia to grant legal protection for those who seek a good death. This was followed by a number of amendments and regulations to expand the accessibility and applicability of palliative care. Moreover, the Patient Right to Autonomy Act (病人自主權利法) of 2016 was considered a great step forward, both conceptually and in practice, so far as hospices are concerned. Through advanced care planning, every patient's own preferences can be secured when encountering clinical situations in which withdrawal of life-sustaining treatment or artificial nutrition are being considered.

At the same time, some experts have reservations about Taiwan's achievements, among whom the most notable is Professor Co-Shi Chantal Chao (趙可式), a pioneering advocate of hospices there. After assisting a governmental investigation into the realization of hospice policies, in an interview in 2020 she revealed a huge gap between institutional efforts and how those visions are actually implemented in clinical settings (Tsao 2020). Only a handful of hospitals meet the standard she sets for quality palliative care; in her experience, many accredited hospice wards are in fact understaffed and do not use their equipment adequately.

As a researcher of science, technology, and society (STS) working on health issues and a certified hospice specialist, I am not surprised by this gap. Like many criticisms of institutionalized health services, Chao's comments are commonplace and can be applied to other hospital departments. What intrigued me in her interview, nonetheless, was a subtle distinction between universal, standardized palliative care and localized, patient-centered care, and the tensions between them. For example, Chao insists that a bathtub is a necessary piece of equipment for every hospice ward, arguing that assisted bathing is a necessary part of hospice practice and should be performed by nursing professionals with proper training. While nobody would deny the benefit that tub bathing could offer in achieving quality care, I am impressed by the notion of considering it a practice that requires professional knowledge and skill. Coining the phrase "nursing for comfort" (*shushi huli*, 舒適護理), Chao creates not only a vague concept redolent of comfort care but a battery of nursing knowledge and practice that is claimed as original and best suited for patients.[3] Found in local textbooks and taught at medical schools in Taiwan, almost every hospice specialist knows the term *shushi huli*, but no exact English equivalent can be found.[4]

This is the starting point of my inquiry. While universal indicators of quality palliative care can be recognized, there are invisible knowledge and practices like *shushi huli* that are culturally bound and professionally performed in Taiwan. In this regard, I find John Law's (2002) concept of "knowledge space" useful. Complicating the meaning of *space* in actor network theory, Law argues that space is not a natural, preexisting container of the social and the material, but is itself a performance, being reified "in a series of sedimented enactments" (95). Objects, according to Law, should be reconsidered in this light. They emerge and travel in different spatial systems, achieving homeomorphism within those systems.

Echoing this, I argue that a hospice ward should be reappraised as a knowledge space where varieties of palliative care are performed by patients, hospice specialists, helpers, and the medical objects among them. This is especially so in Taiwan, where palliative care is dominantly identified as a hospital-based clinical behavior and is strictly regulated. This institutional feature invites criticism of medicalization, as Sharon Kaufman (2003, 2015) has shown in the case of American hospitals, and Taiwan is no exception. While in hospitals the majority of palliative care is performed under shared-care schemes with hospice specialists, hospice wards are not nominal existence, meaning that they offer medical service like other wards do.[5] These wards not only house specialists devoted to end-of-life care but also offer training to other professionals.

In the rest of this chapter, I demonstrate that the hospice ward is not merely a physical location within the modern hospice, such as originated in the United Kingdom in the 1960s and was transferred to Asia in the late 1980s (Goh 2002), but instead a space of epistemic decentralizing where different kinds of knowledge and innovative treatments are developed. These therapeutic practices, I argue, complicate the otherwise standard protocols of the institutional hospice by establishing different kinds of relationship with it.

As part of ongoing research into the transformation of hospice and palliative care in Taiwan, this ethnographic study aims to explore how these therapeutic practices are performed, developed, and settled as part of the caring repertoire in a hospice ward at Hospital VG, one of the leading medical centers of northern Taiwan. The main material comes from my participatory observation as a professional hospice trainee from 2011 to 2013 and as an STS scholar working on the use of technology in palliative care since then. In addition to fieldwork on caring practices, both orthodox and unorthodox, I have conducted interviews with professionals and practitioners who perform them. As a hospice specialist, I have also attended professional seminars and conferences to discover other professionals' perceptions of and reactions to these practices.

Situating Hospice Wards in Taiwan's Health Care System

As in other East Asian states, the idea of introducing modern hospices to Taiwan can be traced back to the early 1980s and to people with religious affiliations.[6] Co-Shi Chao, for example, when she was still a practicing nurse in 1983, was fascinated by the concept of the hospice, which drove her to gain a master's degree in oncology nursing and a PhD with a focus on hospice and palliative care from Case Western Reserve University. During her study abroad, Chao also attended seminars at St. Christopher's

Hospice, the first such modern hospice, founded by Cicely Saunders in 1987. With such professional training, after her return to Taiwan in 1993 Chao quickly joined existing initiatives such as the task force for cancer care at the Kung Tai Catholic Sanipax Socio-Medical Service and Education Foundation (天主教康泰醫療教育基金會), which was followed by the institutionalization of hospital-based hospice and palliative care, notably the foundation of the Hospice Foundation of Taiwan (安寧照顧基金會) in 1990, the Buddhist Lotus Hospice Care Foundation (佛教蓮花基金會) in 1994, the opening of the first hospice ward at Mackay Memorial Hospital in 1990, and the Taiwan Academy of Hospice Palliative Medicine (TAHPM, 台灣安寧緩和醫學學會) in 1999.

The professional movement in hospice and palliative care in Taiwan coincided with the introduction of National Health Insurance (NHI). As early as the second year of its implementation, 1996, the pilot program Hospice Home Care Benefits and Subsidies was launched, which set standards for inpatient hospice and home care. Following legislation in 2000, NHI formally subsidized hospice inpatient care with a per capita, per diem payment scheme.[7] When in 2005 the Department of Health (later the Ministry of Health and Welfare, MHW) launched its National Cancer Control Project 2005–2009, the provision of quality palliative care was included as part of integrated cancer control and care.[8] It became an indicator for cancer care quality assurance measures in the accreditation of hospitals that treated over five hundred new cancer cases in 2008. In 2011, the Joint Commission of Taiwan (JCT, 醫院評鑑暨醫療品質策進會), the official organization for hospital accreditation, started including items on hospice and palliative care in its evaluation list.[9] Covering services in hospice wards, hospice home care, and hospice shared care, these items have been part of hospital accreditation since 2013.

The development of hospice wards in Taiwan should be considered against the backdrop of these regulations. As mentioned earlier, hospice practice in Taiwan began with individual efforts that attempted to provide a different kind of care, and the first organizational move was to carve out a space, that is, a hospice ward, from the regular clinical settings in hospitals. As Dr. Wen-Hao Su, a core member of the hospice movement in Taiwan, recalled, "a specialized ward is the root of the hospice" (Chang 2012). While the early inclusion of hospices in the NHI scheme provided the incentive for hospitals to set up such facilities, from an organizational perspective it also affected the path of hospice care. The former TAHPM secretary, Dr. Chien-An Yao, rightly comments that this is a case of carrot and stick: as hospice and palliative care does not generate profits for hospitals, NHI provides financial support for such services, while the JCT imposes regulations to maintain quality control (Chang 2012).

As a result, what makes Taiwan's hospices distinct from their Western counterparts is the institution. As of 2019, seventy-six hospitals had set up hospice wards, mainly public hospitals and those sponsored by religious organizations, providing 819 beds in total for inpatient palliative care. About 40 percent of these hospice wards were in northern Taiwan, where the capital is located, and the majority had fewer than twenty beds (TAHPM 2019). Professional personnel are another concern in the making of the hospice profession. Not being recognized by the MHW as a medical specialty, the profession instead has been built around qualified hospitals (twenty-six as of 2019), whose wards offer training courses and internship programs. Under the NHI reimbursement scheme, every hospice ward qualifying as an inpatient hospice must have at least one attending physician who has received eighty hours of hospice training (including forty hours of internship) and offer all-year, round-the-clock diagnosis and consultation services. Every bed in a hospice ward should have at least one attending registered nurse who, like the physician, has received eighty hours of hospice training (including twenty hours of internship). In addition to the clinical personnel, every hospice ward needs to have in-house staff (helpers or nurse aides, social workers, nutritionists, and pharmacists) with proper training.[10]

Unlike the Foucauldian imaginary of hospital medicine in which wards present the primary spatialization of clinical specialties (Foucault 1973), in Taiwan the hospice ward is an enclave in a hospital, where professional identity emerges from ward activities—team meetings, ward rounds, reading clubs and seminars, and consultations, among other administrative routines.

Decentralizing Biomedicine in Hospice Practice

While aiming to provide more humane, appropriate care for those who look for a good death, in reality the operation of a hospice ward in Taiwan is like other wards in many aspects. Though a hospice is covered by health insurance, patients are asked to meet any discrepancy if they choose to stay in better surroundings, such as a double room. Also, while patients are allowed to stay in hospice wards for over thirty days, for the majority of them the length of their hospitalization is around fifteen days, just like other medical wards. Nurses take shifts in hospice wards, while attending physicians have work outside of the wards, such as consultations, outpatient clinics, and nonhospice obligations.

The difference between care practice in hospice wards and in regular ones, I would argue, lies in the purpose of treatment and the people who conduct it. Before admission, patients have to sign do-not-resuscitate (DNR) consent forms and agree to admission on the understanding that the goal of their hospitalization is palliative. Treatments are to alleviate the patients' symptoms (e.g., pain) instead of curing them, through, for example, chemotherapies or intensive surgical intervention. In contrast with their patients, the clinical staff, especially nurses, are generally ambitious and proactive. Not only leaders like Co-Shi Chao but many other nursing staff I have spoken to have expressed the differences between themselves and other clinical professionals: they are trained to be proactive in serving on hospice wards and expected to do things both professional and unconventional.

The task of making palliative care both professional and unconventional may look paradoxical. However, it is through the therapeutics introduced by these specialists that I have come to rethink hospice wards as decentralized therapeutic spaces where every therapy is alternative and complementary to each other. East Asia is, in fact, characterized by its multitherapeutic environment. In addition to biomedicine, there are local healing traditions such as Kampo, Korean, and Chinese medicines, which have been modernized and granted a certain professional status and welcomed by local users.[11] Even so, the integration of biomedicine with these major complementary and alternative medicines (CAM) is difficult, not to mention such healing traditions as aromatherapy or art therapy.[12]

In the rest of this section, I demonstrate three such therapeutics used and developed in the hospice wards I have worked on and been associated with as a researcher—aromatherapy and herbal ointment treatment, electronic acupuncture, and qigong induction.[13] Unlike a number of designed therapeutic experiments aiming to combine different knowledge traditions by demonstrating efficacy, therapeutic practices in hospices are fluid, practical, and experience based. Conducted by professionals within the trajectory of hospitalization, as Anselm Strauss would suggest (Weiner et al. 1997), these non-biomedical therapeutics grant on-site experience and allow expertise to develop by being put into practice.

Aromatherapy is the practice of using essential oils for therapeutic benefit. While having been used for centuries and being popular in Western societies, aromatherapy has not been widely applied for medical use in East Asia, where essential oils are thought of as part of a certain lifestyle rather than a therapeutic means. In a hospice ward I observed, aromatherapy was introduced by a senior nurse whom I will call Ms. Chen. Chen first came across aromatherapy when she was at nursing school; even so, it was not until she served on Hospital VG's hospice ward, where there were some donated essential oils, that Chen developed the idea of incorporating these oils into her nursing practice. Already a hospice specialist, when I first met her in 2011 Chen had become certified in aromatherapy by the National Association for Holistic Aromatherapy (a US-based nonprofit organization) and offered training classes when off duty.

For Chen, aromatherapy was new. Not only was there scant literature addressing the symptoms that aromatherapy might help in a hospice setting, but she also had to make sure her patients would accept it in their ward routine. With her experience on the hospice ward, Chen gradually developed a working repertoire based on aromatherapy, but mixing in elements from other alternative medicines. In a treatment table that Chen created for her colleagues, laying out how to apply aromatherapy to hospice patients, I saw notes on the indications for common essential oils such as lavender, tea tree, rosemary, and eucalyptus. Some herbal ointments used in Asian medicines, notably *shiunkou* (紫雲膏), were also included.[14] Instead of treating aromatherapy as complementary to biomedicine, Chen was confident about their combined effects. She demonstrated to me one of the typical cases she had been experimenting with for a while: a patient suffering from a huge bedsore. In addition to the conventional approach of aromatherapy, Chen was able to try her essential oil formula on the wound via wet dressing twice a day, and the bedsore shrank remarkably within a week.

Although often termed *electronic acupuncture* (電子針灸) in Taiwan, the device used in another therapy is more properly referred to as a transcutaneous electrical nerve stimulator (TENS).[15] In biomedicine, TENS is a low-end technology widely adopted along with infrared light stimulators in physical therapy for local relaxation and pain control. However, and bearing in mind the gate control theory proposed by Patrick D. Wall and Ronald Melzack in 1965, hospice specialists in Taiwan also consider TENS as a therapeutic substitute for acupuncture, which by regulation requires a certified professional to perform. Although Hospital VG has a center for traditional medicine, and thus, theoretically, hospitalized patients might receive treatment through interdepartment consultations, in reality hospice specialists tend to arrange TENS treatment for their patients whenever they feel it beneficial. In fact, the TENS machine widely used in hospices in Taiwan was designed by the famous acupuncture researcher Ji Sheng Han (韩济生). Using low-frequency waves (2–5 kHz) like other TENS devices, Han programmed several settings (fifteen minutes per session, mixed stimulations, with various intensities) for different clinical symptoms and has claimed therapeutic effects. In the users' manual he also advises specific acupuncture points, such as *zusanli* (ST36 as in WHO nomenclature), *neiguan* (PC6), *shenshu* (BL23), and *dachangshu* (BL25) for treatment.

It is through TENS that Chinese medicine plays a role in hospice routines. Although TENS devices can freely be purchased by the general public, in hospice wards they are used only by hospice specialists or under their supervision. The clinical indication for using TENS is, as mentioned, rather arbitrary. As a part of palliative care, it is hospice nurses and sometimes physicians who decide when TENS is used on their patients, and on which preset program. In turn, the patients' reactions play an equal role in negotiation over pain control. More than once I have seen patients who have tried out one program on a given day refuse to have a more intensely stimulatory one on another day. Knowing my concerns as to medical compliance, hospice specialists have explained that palliative care is as dynamic as the doctor-patient relationship, in which professionalism cannot stand alone without a patient's satisfaction and comfort.

In my hospice fieldwork, it has been hard to pigeonhole Ms. Guo, a qigong and spiritual master in residence. Instead of simply leading qigong exercises or giving religious talks, she sees herself as a healer of souls who provides energy treatments and consultations. Because she comes to Hospital VG only one afternoon once a week, hospitalized patients have to make appointments with her in advance to secure a treatment slot, usually ten to twenty minutes, which usually includes a group consultation and qi induction. Guo's treatment is precise and effective. In just minutes she seems to see inside the patient and to know their concerns, both biological and social, and interprets them in spiritual terms that are convincing. She is also able effectively to relieve the patient's discomfort with her input of qi. Although a regular healer on the hospice ward, Guo never remembers her patients' names and does not keep any record of her treatment. As she explains, "The patient's condition is always changing; what we can do is only focus on what they need at this moment and give them immediate help."

Guo's presence in the hospital is unconventional. Not only can she find no place for the healing she offers in biomedicine, but there is also no place in Chinese medicine for the therapeutic tradition she belongs to. According to Guo, she possesses a spiritual talent but did not fully discover it until her return from Canada in 1999. Inspired by *zhuyou* (祝由), one of thirteen therapeutic categories in traditional Chinese medicine, Guo started her religious offerings and her treatment of people, first children and then hospice patients, in 2006.[16] For her, the body and the spiritual are inseparable in *zhuyou*, and the hospice ward provides a perfect venue for the revival of such a practice. Although Guo's healing art is too idiosyncratic to be performed by others, she has demonstrated that, moving beyond a treatment paradigm with biomedicine at its core, there are abundant therapeutic possibilities that can complement each other.

Crafting a Decentralized Profession and Its Limits

Thus far I have described hospice wards as a therapeutic and knowledge space in Taiwan, where biomedicine is decentralized by making it one among many alternatives in palliative care. Following on, in this section I further discuss how decentralized palliative care can be achieved in hospices and how an inclusive, therapeutic profession can emerge and be empowered through the people who practice it.

As I wrote previously, hospice wards operate much like hospital wards; the setting can be said to be biomedical. The per capita and per diem payments by the NHI (less than US $245 per day)[17] restricts the use of interventions during hospitalization to the biomedical and does not cover alternative medicines. The cost of non-biomedical services on hospice wards, therefore, is covered by other sources, mostly donations. For example, in Hospital VG, a public hospital, additional palliative services are sponsored by an associated foundation that shares the same space as the hospice ward.

This situation explains why some hospitals in Taiwan can achieve a relatively high quality of care on a budget. On the one hand, they welcome hospice facilities to demonstrate their charitable nature, managing donations to fill in as much as possible the gap between the ideal hospice and the reality of a limited income. On the other, as biomedical institutions, these hospitals would not allow patients to try alternatives during their hospitalization, even if willing to pay out of their own pocket. The hospice specialists, at this point, play a critical role. By their professional judgment they introduce non-biomedical alternatives onto the hospice ward and perform them themselves. Through ward meetings and rounds they also decide and evaluate when and how these alternatives should be used in biomedical treatment trajectories.

Like other clinical specialties, hospice specialists acquire their professional knowledge through authorized textbooks, board-certified training, and continuing education programs. In addition to the TAHPM for physicians, and the Taiwan Association of Hospice Palliative Nursing for nursing professionals, other professional associations and hospitals offer lectures and seminars.[18] To meet standards as professional specialties and allow professionals from other specialties to join, the courses these institutions offer are predominantly biomedical.[19] Nonetheless, the ways to interpret these biomedical practices are flexible. In addition to certain culturally bound teachings on hospice care, such as the "four *daos*" (*daoxie*, *dao'ai, daoqian, daobie*—to thank, to love, to apologize, and to bid farewell), in order to reach a good death or the ultimate goal of family-based hospice care—achieving "the mutual peace of the living and the dead" (生死兩相安)—there are bridging instruments or skills that are considered both biomedical and alternative.

For example, TENS presents as a constructive instrument echoing the notion of interpretative flexibility (Bijker, Hughes, and Pinch 1987). As I mentioned earlier, TENS is a biomedical device for physical therapy but is also seen as falling within the scope of Chinese medicine, which enjoys an equal status with biomedicine in Taiwan. Physicians might not feel comfortable applying TENS to their patients, not sufficiently acquainted as they are with it as a physical therapy, but when I have asked them in seminars about using TENS on hospice patients, they have not been opposed to the idea, as if it were a natural part of Chinese medicine that can be applied to palliative care. It is certain that TENS cannot replace acupuncture, which should be performed by professionals of Chinese medicine. Nonetheless, it is because of its ambiguity that TENS can connect practitioners across specialties and professions.

In the same vein, nursing for comfort translates to clinical skills for both hospice specialists and lay caregivers. As already mentioned, Co-Shi Chao, the inventor of *shushi huli*, insists that it is the foundation of the nursing profession, and nursing for comfort is regularly taught in hospice training programs. Even so, at the same time, Chao invites everyone to learn *shushi huli* and apply it to those who need care. As the textbook she edits (Chao 2014) clearly states, such training is patient centered. It is designed for nurses who have left the profession to become patients' caregivers, and for instructors and students who have forgotten this caring art and cannot acquire it at nursing school. It is also for caregivers who badly need this skill to help their patients. From an STS perspective, *shushi huli* is a "boundary object" (Star and Griesemer 1989). While containing professional content (such as the use of antibiotics and complicated wound care), it is a practical guide that covers everyday situations a hospice patient might encounter. When I discussed this ambiguity with one of the authors of the *shushi huli* textbook, he admitted that the readership in Chao's mind is neither nursing professionals nor laypeople but those who are committed to palliative care and who would become specialists through practicing it.

Chao's patient-centered attitude toward hospice nicely reflects medicine as both science and a source of succor (Collins and Pinch 2005). It also envisions how a decentralized profession might emerge on hospice wards. On the surface, the hospice seems to be a multidisciplinary field, where physicians, nurses, pharmacists, social workers, clinical psychologists, nutritionists, and many other professionals come together to contribute their knowledge. It is true that, under the umbrella institution of the Hospice Foundation of Taiwan, medical and nonmedical members can work together to design a hospice profession for Taiwanese patients. Yet what is equally crucial in shaping this profession, I argue, is in the everyday routine on hospice wards, which, as both therapeutic and knowledge spaces, give everyone involved the freedom to accumulate experiences, to refine skills, to outsource therapeutics for innovations, and eventually to distinguish themselves from other medical professionals. For that, the hospice profession has had to be decentralized since its inception. I remember that when I asked Ms. Chen to write papers on her findings about aromatherapy, she refused: "It is not research," she replied honestly. "I did what I thought good for my patients at that moment and left the recipe for others to develop." In her words, I seem to have found the secret behind quality palliative care in Taiwan.

The reality of crafting a decentralized caring profession, however, is challenging. Despite Taiwan having won a high ranking for its service outcomes, and despite having developed skills to cope with local needs, the MHW has been officially questioned by the Control Yuan over the quality of hospice and palliative care it regulates, and policy reforms have been demanded so as to increase the capacity of services, implement hospital accreditation, and strengthen professional training (Control Yuan 2019). The challenge, as indicated in the investigatory report, lies in the enduring tension between the willingness to acquire additional professional skills and the incentive to practice them. The standard for hospice wards to meet is high; without hospice wards that offer adequate equipment (such as TENS) and administrative support, it would be hard for many hospice specialists to establish their expertise after completing their training. The institutionalization of hospices provides the infrastructure needed; meanwhile, it sets the limits for the profession.

I remember a conversation with an oncologist after a combined meeting with hospice specialists on hospital accreditation for quality of cancer care. The young oncologist was new to hospice and did not quite understand the need to set up a specialized space within the hospital. "We as physicians always do hospices," he said. "What the government should do is not build more hospice wards but offer incentives for other specialists, such as oncologists, to share the service burden." He was right that institutional means—such as hospital accreditations or reimbursement incentives to perform hospice care on regular wards—can increase to a certain degree the service capacity of such care. In fact, it has been the MHW's policy to reach out from hospice wards. Even so, I can see that the hospice movement has been detached from the space its profession begets and is gradually losing the chance to develop its own therapeutic characteristics, which are decentralized and constructive.

Conclusion

Based mainly on intensive fieldwork on a medical center's hospice ward, I have reflected in this chapter on the crafting of the hospice in Taiwan as a decentralized profession. Scholars have commented on the policy issue of budget versus quality in hospice services and on a holistic nature that includes not only biomedicine but also alternative therapeutics. This chapter has characterized hospice wards as therapeutic and knowledge spaces that nurture the profession.

The non-biomedical therapeutics that are introduced to mediate patients' bodily needs, their caregivers' expectations, and medical staff's beliefs regarding hospice approaches should be assessed in the light of epistemic decentralizing. This chapter has argued that recognizing these therapeutic spaces does not mean overturning the biomedical scheme.

While non-biomedical therapies do show some desirable clinical effects, they should be understood as part of the total effort made toward the favorable ending of life.

I further argue that hospice wards consolidate the decentralized nature of the hospice profession. They do so by allowing professionals, based on their backgrounds and expertise, to practice in various clinical situations toward the end of their patients' lives. There do exist disagreements in perspective or in approach among hospice specialists. They occur every day in ward routines. Even so, ways to reach consensus are less hierarchical than on regular wards, where physicians dominate with their biomedical authority. In the morning meeting on a hospice ward, it is the caring team that takes the lead in finding solutions. As hospice specialists, physicians are often listening to others and more open to alternatives if they are helpful. It is through these daily practices that biomedicine is decentralized; this is achieved not by competition with therapeutic alternatives but by discussions and negotiations focusing on the problems at hand.

By way of concluding this chapter, let me remind readers of Eliot Freidson, a pioneering figure in medical sociology. In *Profession of Medicine: A Study of the Sociology of Applied Knowledge*, Freidson ([1970] 1988) rightly calls medics "consulting professions," arguing for their autonomy in a free society. Hospice and palliative care practitioners, I shall say, inherit this spirit but deal with far more complicated situations on the fine boundary between life and death. The concept of the good death is age-old, yet the conventional division of expert knowledge and practice of it may fall short in capturing an emerging profession like the hospice. For this, a vision that features decentralizing knowledge is not just allowing non-biomedicine to have a say in this field; it is also an innovation for a new form of profession awaiting all of us.

Notes

1   In the 2010 study, forty countries were included for comparison. The number increased to eighty in the following study, conducted in 2015.
2   The following content is extracted from three articles by the study team (Bhadelia et al. 2021; Sepulveda et al. 2021; Finkelstein et al. 2021).
3   In hospitalization, *comfort care* means a therapeutic plan that is focused on symptom control, pain relief, and quality of life. It is typically administered for a clinical situation where further medical treatment appears unlikely to change matters and can therefore be said to be equivalent to hospice or palliative care.

4   While the terms *comfort nursing* and *nursing for comfort* exist in English, they refer to a caring situation (such as when a baby sucks on the mother's breast, not out of hunger but to soothe itself to sleep), usually between parent and baby, without a professional's involvement. In her book *The Basics of Comfort Measures* (Chao 2014), however, Chao talks of *shushi huli* as a comprehensive set of caring skills for older patients, with their bodies as its main concern: "the 'basic' of the nursing profession is 'to care for the most fundamental needs of the patient's body.'"

5   In order to promote hospices and improve the quality of care for hospitalized patients in their final days, the Bureau of National Health Insurance introduced a hospice shared-care scheme in 2011. Under this scheme, hospice specialists are notified and join the medical team for patients admitted to regular wards through consultations. In addition to regular treatment, during their hospitalization patients can enjoy features of palliative care, notably discussion of do-not-resuscitate (DNR) directives, among other hospice-style care options, more nuanced pain control, and support for themselves and their families through consultations and family meetings.

6   The term *hospice* was translated into Chinese and introduced by Dr. Chang-Hung Chung (鐘昌宏), an oncologist then at Mackay Memorial Hospital, in 1983. He had learned the concept of hospice when he was a clinical fellow in the United States and had visited the UK to see how it was practiced in medical institutions. For promoting hospice, among other services to international health, Dr. Chung was awarded the Healthcare Model Award (台灣醫療典範獎) by the Taiwan Medical Association in 2012.

7   The enactment of the Hospice Palliative Medical Regulations (安寧緩和醫療條例), which legalized the patient's right to sign a DNR order. These regulations were amended in 2002, 2011, 2013, and 2021 to allow the withdrawal of life-sustaining devices for terminally ill patients if predetermined by themselves or agreed by their family members.

8   In the Cancer Control Act of 2003, the availability of hospice services for terminal cancer patients was listed as one of five tasks for cancer control.

9   Established in 1999 and funded by the Ministry of Health and Welfare, the Taiwan Hospital Association, the Taiwan Nongovernmental Hospitals and Clinics Association, and the Taiwan Medical Association, the JCT was commissioned to promote, execute, and certify the nation's health care quality policies via accreditations, aiming to enhance the management of health care organizations.

10 It is a regulation that for every three hospice beds there should be one helper or volunteer with appropriate education in end-of-life care. Every hospice ward should have at least one social worker who has completed one hundred hours of hospice training (including forty of internship). In addition to nutritionists and pharmacists, hospice wards are advised to have in-house clinical psychologists, physical and occupational therapists, and spiritual and religious counselors.

11 For the recent utilization of these healing traditions in Taiwan, Japan, and Korea, see studies conducted by an international team led by Ichiro Tsutani, Myeong Soo Lee, and Fang-Pey Chen (Lee et al. 2018; Motoo et al. 2019; Huang et al. 2019).

12 According to the US National Center for Complementary and Alternative Medicine (NCCAM; 2000), CAM is defined as "a group of diverse medical and health care systems, practices, and products that are not presently considered to be part of conventional medicine" and can be recognized in five categories: alternative medical systems (such as Kampo or Ayurveda), mind-body interventions (such as naturopathy), biologically based treatments (such as herbal or dietary), manipulative and body-based methods (such as chiropractic), and energy therapies (such as qigong, Reiki, prana, or therapeutic touch).

13 Other therapeutic alternatives are used in hospice wards, such as art therapy, massage therapy, psychological or spiritual consultation, or *shushi huli*. I discuss the use of these services later in this chapter.

14 Said to have been created by the famous eighteenth-century physician Seishu Hanaoka, *shiunkou* is a traditional antibacterial and anti-inflammatory remedy in Asian medicine, widely used in treating eczema, mosquito bites, and small wounds, among other external uses.

15 Unlike TENS, electronic acupuncture requires the insertion of acupuncture needles into the patient's body, through which electronic stimulation is applied.

16 *Zhuyou* can be understood as shamanic healing and presents as one of the prototypes in the making of Chinese medicine. Although it has been considered unorthodox compared to herbal remedies and acupuncture, *zhuyou* is still a valid repertoire in Chinese medicine and is mentioned in textbooks and research literature.

17 Under the global budget scheme, payment is counted as reimbursement units, whose value floats upon the total number of units spent in a fiscal year.

18 Such as the Taiwan Society of Cancer Palliative Medicine (台灣癌症安寧緩和醫學會), the Taiwan Community Hospital Association (台灣社區醫院協會), or the Taiwan Association for Medical Continuing Education and Promotion (台灣醫療繼續教育推廣學會).

19 Training courses for physicians and nurses are required to include the following eight themes: introduction to hospice and palliative care, symptom control and care for comfort, psychosocial and spiritual care of hospice patients and their family members, grief counseling for hospice patients and their family members, legal and ethical issues in hospices, issues surrounding communication between the patient and the hospice team, introduction to hospice services (on the hospice ward, shared care in hospitals, and home care), and internship.

References

Bhadelia, Afsan, Leslie E. Oldfield, Jennifer L. Cruz, Ratna Singh, and Eric A Finkelstein. 2021. "Identifying Core Domains to Assess the 'Quality of Death': A Scoping Review." *Journal of Pain and Symptom Management* 63 (4): e365–86. https://doi.org/10.1016/j.jpainsymman.2021.11.015.

Bijker, Wiebe E., Thomas Parke Hughes, and Trevor Pinch, eds. 1987. *The Social Constructions of Technological Systems*. Cambridge, MA: MIT Press.

Chang, Ya-Wen. 2012. "Nine Items on Hospice and Palliative Care Will Be Included in the Hospital Accreditation; Specialists Look Forward to Further Regulations" [in Chinese]. Hospice Foundation of Taiwan. https://www.hospice.org.tw/content/1558.

Chao, Co-Shi Chantal. 2014. *Zhaohu jiben gong—Zhao Keshi jiaoshou shizuo jiaoxue* [照護基本功— 趙可式教授實作教學, The basics of comfort measures]. Taipei: Farseeing Publishing Group.

Collins, Harry, and Trevor Pinch. 2005. *Dr. Golem: How to Think about Medicine*. Chicago: University of Chicago Press.

Control Yuan. 2019. "Investigation Report: A Case to Rectify the Ministry of Health and Welfare on Hospice and Palliative Care" [in Chinese]. https://www.cy.gov.tw/CyBsBoxContent.aspx?n=133&s=6866.

Duke-NUS Medical School. 2022. "Cross Country Comparison of Expert Assessments of the Quality of Death and Dying 2021." https://www.duke-nus.edu.sg/lcpc/quality-of-death/.

Finkelstein, Eric A., Afsan Bhadelia, Cynthia Goh, Drishti Baid, Ratna Singh, Sushma Bhatnagar, and Stephen R. Connor. 2021. "Cross Country Comparison of Expert Assessments of the Quality of Death and Dying." *Journal of Pain and Symptom Management* 63 (4): e419–29. https://doi.org/10.1016/j.jpainsymman.2021.12.015.

Foucault, Michel. 1973. *The Birth of the Clinic: An Archaeology of Medical Perception*. London: Tavistock.

Freidson, Eliot. (1970) 1988. *Profession of Medicine: A Study of the Sociology of Applied Knowledge*. Chicago: University of Chicago Press.

Goh, Cynthia R. 2002. "The Asia Pacific Hospice Palliative Care Network: A Network for Individuals and Organizations." *Journal of Pain and Symptom Management* 24 (2): 128–33.

Huang, Ching-Wen, Diem Ngoc Hong Tran, Tsai-Feng Li, Yui Sasaki, Ju Ah Lee, Myeong Soo Lee, Ichiro Arai, Yoshiharu Motoo, Keiko Yukawa, Kiichiro Tsutani, Seong-Gyu Ko, Shinn-Jang Hwang, and Fang-Pey Chen. 2019. "The Utilization of Complementary and Alternative Medicine in Taiwan: An Internet Survey Using an Adapted Version of the International Questionnaire (I-CAM-Q)." *Journal of the Chinese Medical Association* 82 (8): 665–71.

Kaufman, Sharon. 2003. *And a Time to Die: How American Hospitals Shape the End of Life*. Chicago: University of Chicago Press.

Kaufman, Sharon. 2015. *Ordinary Medicine: Extraordinary Treatments, Longer Lives, and Where to Draw the Line.* Durham, NC: Duke University Press.

Law, John. 2002. "Objects and Spaces." *Theory, Culture and Society* 19 (5/6): 91–105.

Lee, Ju Ah, Yui Sasaki, Ichiro Arai, Ho-Yeon Go, Sunju Park, Keiko Yukawa, Yun Kung Nam, Seong-Gyu Ko, Yoshiharu Motoo, Kiichiro Tsutani, and Myeong Soo Lee. 2018. "An Assessment of the Use of Complementary and Alternative Medicine by Korean People Using an Adapted Version of the Standardized International Questionnaire (I-CAM-QK): A Cross-Sectional Study of an Internet Survey." BMC *Complementary and Alternative Medicine* 18 (1): 238. https://doi.org/10.1186/s12906-018-2294-6.

Motoo, Yoshiharu, Keiko Yukawa, Ichiro Arai, Kazuho Hisamura, and Kiichiro Tsutani. 2019. "Use of Complementary and Alternative Medicine in Japan: A Cross-Sectional Internet Survey Using the Japanese Version of the International Complementary and Alternative Medicine Questionnaire." JMA *Journal* 4 2 (1): 35–46. https://doi.org/10.31662/jmaj.2018-0044.

National Center for Complementary and Alternative Medicine. 2000. *Expanding Horizons of Healthcare: Five-Year Strategic Plan 2001–2005.* NIH Publication No. 01-5001. Washington, DC: US Department of Health and Human Services.

Sepulveda, Juan Marcos Gonzalez, Drishti Baid, F. Reed Johnson, and Eric A. Finkelstein. 2021. "What Is a Good Death? A Choice Experiment on Care Indicators for Patients at End of Life." *Journal of Pain and Symptom Management* 63 (4): 457–67. https://doi.org/10.1016/j.jpainsymman.2021.11.005.

Star, Susan Lee, and James R. Griesemer. 1989. "Institutional Ecology, 'Translations' and Boundary Objects: Amateurs and Professionals in Berkeley's Museum of Vertebrate Zoology, 1907–39." *Social Studies of Science* 19 (4): 387–420.

TAHPM. 2019. *Splendid Twenty Years in Hospice and Sustaining It for the Future: Celebrating the Twentieth Anniversary of Taiwan Academy of Hospice Palliative Medicine.* Taipei: Taiwan Academy of Hospice Palliative Medicine (TAHPM).

Tsao, Fu-nien. 2020. "Mother of Hospices Turns to a Patient Admitted to Hospice: Co-Shi Chao Admits That Taiwan Replies 'Model Hospitals' to Achieve High Ranking in Hospice" [in Chinese]. *The Reporter*, December 20, 2020. https://www.twreporter.org/a/good-death-myth-behind-the-asia-champion-of-the-quality-of-death-index.

Weiner, Carolyn L., Shizuko Fagerhaugh, Barbara Suczek, and Anselm L. Strauss. 1997. *Social Organization of Medical Work.* Reprint, Piscataway, NJ: Transaction.

NINE / RONALD CANCINO, CRISTINA FLORES,
ELÍAS BARTICEVIC, AND HEBE VESSURI

Decentered Scientific Agendas and Decentralized Actors and Capacities in Patagonian Science

Despite abundant rhetoric to the contrary, the deprioritization of the local is a real problem in today's world. Lack of a serious national engagement with the local domain is often detrimental to the production of effective knowledge, given the absence of healthy intellectual debates and considered deliberations. The chain of subordination extends through a stratified scheme that hinders recognition of the lower echelons in the geographical scale concerning a culture of self-validation, self-assertion, confidence, and independent judgment. The existing practice and logic by which expert knowledge is recognized often has little to do with the merit and intellectual standing of the persons in question. A common reaction to discourses coming from distant provincial territories is to either ignore, minimize, or dismiss them. Scientists working in such regions must deal with this when they wish to prove the value and applicability of their local work.

The possibility of building a relatively autonomous or better-articulated science agenda for a local territory often ends up embroiled in controversy if there is no adequate theorizing of a shared common history and interests. In the past, the divide between pure and applied research usually came down on the side of the former. Today, however, the deep entanglement of research with economic and technological issues is widely recognized, and, in such a context, the definition of a piece of work as being of local interest tends to become the basis for seeing it as parochial and limited, not transcending the immediate context.

Different logics and eventual synergies between social and political practices seeking to empower local social actors formally aim at building and consolidating institutional and organizational capabilities, although not necessarily addressing their empowerment. These include remote provincial groups, marginal communities, and heterogeneous actors, be they scientists, businessmen, civil society, indigenous, peasants, or other people. The mention-and-inclusion strategy often found in discourses at the national level of decision-making produces the effect that merely lip service is paid to critical voices that continue to remain outside and to be irrelevant. The mode of inclusion is problematic. The empirical discussion and presence of local voices in this argument often lead to closure—both intellectual and structural—and thwart attempts at further deliberation. The issues are deemed to have been raised, highlighted, addressed, and resolved (Connell 2007). This manner of addressing local claims does not create further spaces for assessing counterarguments or responding to them, a situation that seems counterproductive to a more equitable national agenda of development.

The limits of state action are porous, since they are the result of permanent processes of challenge, restoration, and relegitimation carried out by people, groups, and state and nonstate institutions. Decentralizing the state here means to cease looking at the state from its core: the capital cities and the agencies of the national executive power. One of the levels of decentralization is expressed in the epistemological—but also political—wish to show not only the power and majesty of the state but to offer a representation of what it could not or can do, of the places to which it did not reach but where it could intervene successfully.

Decentralization attempts to reverse rapidly and sometimes traumatically the results of long historical processes by means of the transfer of resources, attributions, and power in general from the top of the state toward the base of the same state or society. Behind the profuse common usage of some terms (decentralization, local participation, efficiency, etc.) there are proposals of very diverse contents that remain opaque behind these terms. Decentralizing may mean incorporating into the assemblages, with equal decision-making powers, actors that had previously been marginalized, ignored, underestimated, or directly eliminated. In general, in Latin American literature, two kinds of proposals can be identified: neoliberal and democratizing. In both, the state becomes an abstract place where the causes of all evils are located, without a real, objective analytic foundation to determine which problem is the result of which aspect of historically concrete policy displayed by a particular state.

In this chapter, we explore how the notions of decentralization and decentering operate through examples of specific, grounded, and situated authoritative cognitive practices, the circulation and use of knowledges, their potential, and limitations, in the southern tip of the South American continent, in two neighboring countries, Chile and Argentina. We review the ways in which state agencies and officials located in social and/or geographically peripheral areas with respect to the metropolitan centers of both countries, replicate (or not) what happens at the national levels or innovate creatively. The aim is to look at them as spheres of relationships and identities in which the porosity and susceptibility to change in social spheres, as well as personalized commitments and exchanges between individuals and groups, operate as effective determinants of state activities in progress. In fact, it is about decomposing the categories *Argentine state* and *Chilean state* to explore the multiplicity of rationales, interests, and intentions present in the creation and performance of the agencies and the subjects that compose them at the regional level. The two national political architectures impose different perceptions, a formally centralized one in Chile and a decentralized one in Argentina.

Although we do not deal explicitly with the historical-cultural logic of territorial occupation, we cannot ignore it. Indigenous extermination generated a historical-cultural debt that is difficult to avoid when indigenous demands are reactivated in contemporary conditions, providing a geopolitical reason for population concentration in specific places in a vast territory. The dynamics and main characteristics of institutional architectures result in similar local processes coexisting with highly translocal ones in a shared panorama of socioenvironmental conflicts associated with mining, aquaculture, and fire.

In Argentina, the Patagonian space is politically and administratively divided into five provinces that from north to south are Neuquén, Río Negro, Chubut, Santa Cruz, and a fifth one that encompasses Tierra del Fuego, Antarctica (the Argentine claimed territory), and the South Atlantic Islands. This configuration affords distinctive features to the region: in provinces, in terms of administrative territorial organization; and on account of its autonomy and autarky in political terms, since the provinces administer their finances and can dictate their own laws, although they cannot contradict any articles of the National Constitution. Some illustrative cases show the heterogeneity of treatment, especially regarding the policy instruments related to the management of natural resources and events representing loss of biodiversity. This is also reflected in decision-making, agreement, or controversy relative to the viewpoints of the governmental administration, and concerning citizens who not infrequently use other prisms to build perceptions and understandings about resource management.

The southernmost macroregion of Chile is made up of the Aysén, Magallanes, and Antarctic regions. This space has historically been economically small. Its isolation and insularity have had a relevant role in foreign policy matters, but also as an incentive to private investment in extractive activities, generating a territorial integration that would lead to serious difficulties for the territories (Román 2020). At present, they are not even 2 percent of the national gross domestic product (GDP). The low relative weight of this macroregion in Chile's total production is fundamentally associated with the small population settled in the area, in addition to the low number of workers. Both Aysén and Magallanes have experienced a structural change at the sectoral level, with a reduction in the importance of agriculture, forestry, and livestock. On the other hand, particularly in Magallanes, manufacturing, construction, communications, transportation, and financial services have grown significantly. Along with this and associated with the strong presence of the state, both due to the relevance of the public administration, as well as investment and subsidies, there has been a sustained improvement in social conditions, health, education, poverty, inequality, and the human development index. Currently, in this territory a new macroregional science, technology, and innovation policy is being built in which scientific actors interact with political, social, and private actors, to define short-, medium-, and long-term priorities.[1]

Three significant phenomena—mining (extractivism), salmon culture (pollution), and forest fires (loss of biodiversity linked to logging and conservation)—and the awareness of local and global socioenvironmental effects result in a highly complex panorama that helps us understand the dilemmas in decentering and decentralizing research agendas. The general process could be defined as follows. In view of extractive activities that have negative impacts on the environment of Patagonia's binational territory, scientists in both countries tend to support research lines that are strongly consistent with the particularities of their territories. Socioenvironmental conflicts and sociotechnical controversies activate citizens, government, and scientists to build decentered research agendas that articulate the territorial specificity with the concerns of global agendas around climate change and its effects, conditions, and causes in the relationship between microterritories and global macrotransformations. In what follows, we investigate these processes around mining, salmon farming, and Patagonian fires.

Mining: Extractivism and National Policies

In Chile, mining expansion relies on an extractive model based on a concession system established at the constitutional level and regulated by the Mining Code and other provisions not contradictory to the previous ones (Ministerio de Minería 1982). The main aspect causing conflicts is the location of the mining operations, generally in conservation areas, parks, and national monuments, due to their huge energy demand and the escalation of impacts they unleash (Herrera 2021). The main impact occurs in connection with protected areas, conservation sites, wetlands, and glaciers, and in relation to the landscape and tourist value of the territory. In these conflicts, networks of nongovernmental organizations (NGOs) and local social organizations are activated, such as Agrupación Aysén Reserva de Vida, Agrupación Puro Ibáñez, Agrupación Antukulef, Agrupación de Turismo Chile Chico, union organizations such as the Medical College of Chile, the Association of Cattlemen, and Chacareros General Carrera Lake.

Around the projects in Aysén and Magallanes, environmental organizations such as Fundación Terram, Greenpeace, and National Geographic Pristine Seas; indigenous communities; scientists; and local social organizations, such as Alianza Carbón, Frente Ecológico Austral Defensa, Civil Society for Climate Action, and the Last Hope, are articulated in citizen networks that generate new requirements for science and national and regional politics.

In Argentina, the mining issue involves strong social criticism and exposure of the large transnational mining companies given the economic-technological characteristics that define the type of exploitation. Metalliferous exploitation represents one of the main activities led by foreign companies, taking advantage of the flexibility of regulations in the country to bring in their high-profit businesses. Technical change due to the transition from underground mining to open-pit mining using leaching or flotation processing techniques to extract the mineral sought in the rock is paramount (Machado et al. 2011). This led several localities in Argentina to mobilize in opposition to megamining and to policies of high negative impact on the environment, such as hazardous waste repositories in Patagonia. This led municipalities to declare "No nuclear," "No to mining" (against megamining; Wagner 2016). Since the early 1990s, social activism supported by NGOs opposing projects that lack environmental impact studies and consume inordinate amounts of water and energy resources has intensified (Fundación Vida Silvestre Argentina 2010). This issue is a current and relevant conflict for Patagonia, since freshwater sources are seriously affected due to the type of extraction technology used. Ecosystems suffer irreversible deterioration, so the local population has expressed its rejection of these activities, especially in mountain range environments. In many cases, the installation of new ventures has been prevented by citizen action.

Salmon Farming and Contamination

The General Law of Fisheries and Aquaculture of Chile (Ministerio de Economía, Fomento y Reconstrucción 1992) delegates to the Undersecretariat of Fisheries (Subpesca) the power of assigning aquaculture concessions and the definition of benthic management and exploitation areas. The role of Subpesca (dependent on the Ministry of Economy, Development and Tourism) is to respond to private requests for concessions. In this institutional framework, aquaculture activity, and especially the economically relevant salmon farming, has grown in the country amid continuous socioenvironmental questioning, the articulation of technological innovation capacities in its techno-productive chains, and the activation of controversies around the causes and effects of marine pollution, the infectious salmon virus (ISA), the response capacity of the national scientific community, the flourishing of harmful algae, and territorial displacement toward the southernmost part of the country. By 2021 there were some 1,158 concessions in the Aysén and Magallanes regions, 791 in Aysén and 367 in Magallanes (Herrera 2021).

Socioenvironmental problems are always present, especially linked to the blooming of harmful algae associated with the dumping of dead salmon in the sea (the causal relationship between both phenomena is a matter of controversy between private, social, and scientific actors). The ISA virus crisis between 2007 and 2010 (which resulted in a 20 percent reduction in exports and the loss of 50,000 jobs) set a precedent, exposing the flaws in regulation (companies hide information about the existence of viruses in their plants; the feeling of ad-porta crisis spreads like a rumor among company managers) and the lack of local scientific capacity to face the problem. In 2016, there was a strong crisis on the Isla Grande of Chiloé, because of the massive dumping of salmon, and a subsequent crisis in harmful algae blooms, which unleashed unprecedented social and political conflict. Citizens blocked access to the island and protests and violence broke out, which resulted in questioning not only the salmon farming model but also the Chilean development model (Herrera 2021). This provoked demands for science to account for the causes of the phenomenon and provide possible solutions.

The main current conflicts around aquaculture in Chile's Austral territory are in Aysén and Magallanes. In Aysén, there is a strong conflict over the blooming of harmful algae in the Jacaf and Puyuhuapi canals, which are protected areas. Five thousand tons of salmon have died in farms, impacting biodiversity and triggering controversy over the causes and effects of salmon dumping and algal blooms. In Magallanes, the most recent case since June 2019 is known as Salmonleaks, involving the NovaAustral company. The Norwegian company adulterates figures on salmon mortality, and it hides fish mortality on the seabed under heavy machinery. Damage to the environment is confirmed in the Agostini National Park, and a new conflict has been triggered regarding the relocation of the farming centers to the Kaweskar National Reserve. The social actors forming networks around these conflicts are global, national, and local. Greenpeace and National Geographic Pristine Seas, as well as the national organization Terram Foundation, together with indigenous communities, scientists, and environmental organizations, mobilize citizens and the scientific world. The Argentine ban on salmon farming in Tierra del Fuego has strongly impacted the debate in Chile, generating a binational articulation around the same socioenvironmental dilemmas and a debate on the freezing of concessions on the Chilean side.

Argentina has comparative advantages for aquaculture, with vast natural resources: excellent temperature conditions, broad availability of sweet water, and extensive areas of sea culture. While there is wide availability of raw materials for aquaculture food (corn, wheat, and soybeans), and the country has knowledge in universities and organizations devoted to aquaculture (Argentine Investment and International Trade Agency/Agencia Argentina de Inversiones y Comercio Internacional; AAICI 2020), it remains a marginal producer not free from problems for foreign investment.

Salmon farming is one of the most controversial issues in southern Argentina. The argument is that the introduction of exotic species such as salmon and the consequences generated by intensive activity in the use of water in uncontrolled environments can have adverse effects, damaging both the water quality and the native species (Carciofi and Rossi 2021). The early introduction of salmonids at the beginning of the twentieth century in continental lakes resulted in the rapid colonization of the environment and the reduction of native species (Sanguinetti et al. 2014). Aquaculture production in the region does not have a fundamental productive role, which, congruent with the business data of foreign companies that accumulate added value outside the region, leaves a rather small balance for the local population. A high percentage of the activity in this region is in private hands of different origins. In the survey of the Council for Structural Change dependent on the National Government of Argentina conducted in 2021, seven private companies devoted to trout production are mentioned for this region, with over 80 percent of the total profits not remaining in the territory (Carciofi and Rossi 2021).

Fire and Megafire Risks to Biodiversity and Conservation

Anthropogenic causes are currently considered the main ignition factor in terrestrial ecosystems. Specialists recommend distinguishing between fire and wildfire.[2] Fire is associated with the evolution of the human species, which has used fire since prehistoric times in agricultural activities and the formation of cultural landscapes (Mouillot and Field 2005). Fire has responded to different intentions, becoming more complex over time. *Wildfire*, or *bush fire*, is defined as the free and unscheduled spread of fire over vegetation. Current malicious fires to a large extent correspond to dark economic interests.

Chile is a country with a robust fire monitoring system, although it may not sufficiently monitor impacts. In recent years, it has shown a sustained increase in the occurrence of wildfires, reaching a historic record in the 2016–17 season with some 600,000 hectares burned (González et al. 2020). The wildfires that occurred in January 2017 in the central region of Chile, the most destructive in the country's history, exemplified how complex interactions between different atmospheric phenomena and conditions (drought, high temperatures, persistence of high pressure in subtropical latitudes of the Pacific and low pressure on the Antarctic periphery), exacerbated by global climate change processes, can cause wildfires of extraordinary magnitude, exceeding the conventional forecasts and capacities historically observed in wildfire prevention and control (Bilbao et al. 2020). Following the 2017 firestorm, there was an intense debate about the causes that produced this disaster, leading policymakers to reconsider Chilean forestry development policies.[3] National authorities began the process of transforming the National Forestry Association (CONAF 2017) into a public service to strengthen its control capacity. The major challenge for environmental authorities is to overcome an eminently sectoral and compartmentalized state administrative structure and include an integrated environmental management that requires high levels of institutional coordination and cooperation (Gobierno de Chile 2017).

The fire problem in Chile has also been strongly linked to criticism of the expansion of forest exploitation. Mapuche communities have historically been linked to the collection of the fruit of the Chilean pine (*Araucaria araucana*), the exploitation of dead *alerce* trees (*Fitzroya cupressoides*), and the collection of plants and herbs from native forests. The problem here is that the forest industry is in traditional Mapuche territory, placing the demand for the recovery of indigenous lands at the center of the debate. When recovering territories, indigenous communities activate productive processes of the native forest, thus generating a strong conflict with the state and forestry companies (owned by the wealthiest families in the country). Communities are accused of being responsible for the illegal cutting and burning of forests, while the communities argue that forest fires are industrial strategies for the extension of the territory subject to commercial forest exploitation.

In northern Neuquén, Río Negro, and Chubut provinces of Argentine Patagonia, fast-growing exotic conifer plantation establishments began in the late 1990s, at a rate of 1,000 to 1,500 hectares per year (Schlichter and Laclau 1998; Rafaele, Núñez, and Relva 2015). During this period, large fires occurred in Nahuel Huapi National Park in Bariloche (northeast of Argentine Patagonia). The relationship between the growing incidence of wildfire, demographics, and accelerated urbanization processes can be illustrated in the case of the city of Bariloche. In this region, data show that fires increased from less than 50 to more than 1,800 per season (5,900 percent), especially in the urban-forest interface. While residential development has been stronger toward the steppe-forest ecotone, the humid and mesic forests of the Nothofagus forest region have also been subjected to greater use for both residence and recreation in this large tourist destination, especially since the early 1990s.

Considering the legal instruments that regulate agencies on firefighting and prevention in the different jurisdictions, the provinces, especially the most affected ones (Neuquén and Río Negro), have taken early measures on their territories, seeking to regularize the action of combat and penalties applied to those responsible for the fires. Integrating the voices of different organizations, including governmental and nongovernmental, provincial bodies, and national entities, all these regulations have in common mainly the protection of native forests in their regions.[4] About decentralization, it is thus observed, in the case of northern Argentine Patagonia, that the provincial jurisdictions make their own decisions concerning the regulation of fires. In the 1990s these laws were the foundation for the main national federal regulations for fire management.[5]

Scientific Capacities

In the context of economic transformations and conflicts in southern Chile, scientific capacities present several characteristics. First, they are small-scale, representing a small technological scientific microsystem marked by the presence of few local higher education institutions (a regional public university in each region—the University of Aysén since 2015 and the University of Magallanes since 1961). Together, undergraduate students represent 0.7 percent of the national total enrolled in the higher education system (Servicio de Información de Educación Superior—SIES; Mineduc 2023). The main academic personnel are located at universities and in several research centers financed with public funds, which are predominantly central, directed at the level of ministries and/or centralized public agencies. University funding comes from the Ministry of Education, while competitive funding for scientific research comes from the National Research and Development Agency. Rules and incentives are thus defined at the central levels of state administration, a characteristic of the Chilean political model, based on political centralization and neoliberal logic of competition for science resources.

When analyzing capacities in terms of specific disciplines in the Aysén and Magallanes regions, the sciences oriented toward understanding ecosystem dynamics and anthropic action predominate, involving research capabilities in marine biology, ecology, marine ecosystems, oceanography, plant science, environmental science, zoology, anthropology, physical geography, meteorology, evolutionary biology, and molecular biology. They are about the natural imprint and the specificity of the territory, and the relationship between fragility and the unique natural condition, as well as the possibility of observing with greater resolution changes and impacts of anthropic action of local and global nature (Cancino et al. 2021). Decentered research lines oriented toward the territory link local research with global agendas, while local research allows the observation of causes, conditioning factors, and effects of associated transformations, such as climate change. At present, a scientific consortium between the University of Magallanes, the University of Aysén, and the Chilean Antarctic Institute together with the Patagonia Center for Research in Ecosystems aims at building a roadmap for Patagonian science, that is, a specific way of producing, circulating, and/ or using knowledge in the search for scientific priorities articulated around climate change and research on marine ecosystems, terrestrial human settlement, sustainable management, health, and biological resources.[6]

How, then, can a small system, constrained by centralized institutional designs that leave few degrees of freedom to direct scientific capacities, without significant resources for regional decision-making, manage to articulate the construction of its own research agendas, collecting specificity and biocultural imprints and articulating local/global concerns? And how can this generate a dynamic of decentralization of research lines and agendas? The depth of the crisis led to the articulation—as a prelude to decentralization processes—of not only the scientific disciplines associated with aquaculture, deforestation, and mining, but also the research from multiple disciplines about local and national development problems.

The population in Argentine Patagonia is much smaller than in the rest of the country (from 1.59 percent in the case of Río Negro to 0.68 percent in Tierra del Fuego and dependencies); accordingly, its scientific capacity is much more restricted. There has been an effort to endow the region with universities and research institutes. There are thirteen national universities of modest to weak strength, and several research institutes of varying complexity. The decentralization of the National Council for Scientific and Technical Research / Consejo Nacional de Investigaciones Científicas y Técnicas (CONICET) was a long process, as each scientific-technological center tried with varying success to generate its own regional agenda based on strategic areas. In terms of R&D expenditure, Río Negro has a surprising 3.54 percent share of national expenditure, configuring a significant cluster of research activity in Bariloche involving high-tech companies, institutions, and scientific and technological activities. It registers an average of 13.82

full-time researchers for every thousand members of the economically active population. Since the national average is 1.67, this would be one of the cities with the highest territorial concentration of researchers in the country. The city encompasses a considerable diversity of high-technology areas (nuclear medicine, renewable energy, nuclear energy, satellites, aerospace science, nanotechnology, material science) and it is today a center of scientific-technological production in the country with international influence. Chubut follows, with 1.47 percent, mostly concentrated around Comodoro Rivadavia, and the rest have much smaller figures (Neuquén, 0.52 percent; Santa Cruz, 0.40 percent; and Tierra del Fuego, 0.37 percent; Subsecretaría de Estudios y Prospectiva 2019).

There is a desired, and not always realized, link between scientific-technological production, economic development, and protected areas. Two major areas stand out: natural sciences and politics, considering the problems of the provinces. Earth and environmental sciences prevail in specific academic nuclei, especially around issues related to climate change, geology, and oceanography. The social sciences, humanities, and the arts have grown modestly as universities and research centers were created, addressing problems inherent to the region, full of migration and indigenous peoples' issues. Finally, the exact sciences (science, technology, engineering, and mathematics, STEM) have also been present in the region since midcentury.

When analyzing cognitive capacities in science and technology, then, we find parallelisms between Argentina and Chile. In both cases there is a decentralized capacity that, through processes of activation of conflicts, citizen mobilization, and political decisions, contributes to deepen the decentralization of the research agendas.

Discussion

We have illustrated the territorial nature of mining, salmon farming, and fire implications of lumber activities in the southernmost regions of the two countries. Somehow territorializations are "enacted into being" (Law 2007) by policies, legal frameworks, concessions, management plans, and so on. These processes are highly translocal. In the two countries, these activities are pushing for access to territories that are classified as valuable and vulnerable and as protected in the countries' management plans. Resource availability is seen in the two countries as a pressing matter, as several arguments strategically work to enroll new territories, for space constitutes a vital infrastructure for these activities. The way circumstances shape the national debates regarding the possibility of intervening in environmentally sensitive areas presents several similarities between the cases. In differentiated ways in Chile and Argentina, mining, salmon farming, and lumber-extracting activities are emblems of the development model. Such highly profitable activities are deployed in territories of high environmental vulnerability and unique biodiversity and within the framework of weak institutions and a system of flexible incentives and regulations. They negatively impact the territories, generating citizen mobilization. The arguments officially employed to legitimize and politically anchor the possibility of operations in areas that have been classified as vulnerable in research documents and management plans are tied to the importance of preserving the country's role as a stable international producer.

The construction of local agendas currently activates a decentering dynamic, gathering partial proposals that express the specificities associated with biodiversity and natural-cultural heritage. The small-scale local scientific capacities strive to give continuity to lines of research articulating concerns about the environmental and patrimonial damage that the types of local and national development impose. A fledgling responsible southern science is particularly mobilized currently in Chile, in a transition to decentralization mechanisms activated by three processes, or what we call turns: (1) a turn toward social demands after the social uprising; (2) the installation of the global climate change agenda as fundamental in local political events, improving the possibilities of local authorities to enhance science and technology capacities; and (3) political decentralization in matters of regional development and especially in science, technology, and innovation.

The Socioterritorial Turn in Scientific Concerns

The presence of conflicts or sociotechnical controversies around extractive activities involves citizen mobilization and the requirement for scientific explanation of the causes and effects of conflicts. Both mechanisms make visible and place citizens' demands at the center of debates (de Sousa Santos 2018). These can be picked up by political authorities, such as the prohibition of salmon farming in southern fjords and channels in Argentina and the current debate on territorial control of aquaculture expansion in Chile. From the point of view of scientific agendas, this constitutes an incentive for the realization of local agendas for the analysis of specific territorial transformations, the relationship between ecosystems and human activity, and the role of social factors in the generation and management of disasters such as wildfires or algal blooms.

In the case of Chile, two milestones are central in this regard: the so-called battle of Aysén of 2012 articulated social and socioenvironmental demands, activated the generation of improvements in local policies, and resulted in the realization of the regional aspiration for its own university, the University of Aysén. Social demand resulted in decentering research lines and activated the decentralization of scientific capacities. A second milestone was the social explosion since October 2019. Social activism, and the reform of the political system that led to the formation of a Constitutional Convention, included science and knowledge as an axis of discussion. It became a matter of Chilean society defining the role and institutional design of science and technology. Unfortunately, the reforms in these fields were not approved by the citizens.

In Argentina, in the 2012–15 period the science and technology system was modified in the context of some national research institutions, by transferring skills and producing collaborative projects as well as relocalizing resources in faraway territories to study local problems. This policy led to the generation of new organizations due to the local presence of some research groups from CONICET. Also noticeable in recent decades was the creation and promotion of a national university in each province in the region, which implied the settlement of researchers and academic activities seeking to generate local critical mass. Still in progress, this initiative begins to inspire specific agendas for local problems and needs.

Historically, Buenos Aires managed the fate of the country's science, promoting ideas such as federalizing science, doing science in the periphery, and so on, while a realistic political commitment accompanied by the right levels of funding was often lacking. Projects with flagrantly insufficient funding have been set in place, justifying them with the argument that they would be carried out in concert with local universities, ignoring that those in the south have a very modest, even precarious, existence, being utterly unable to support the required levels of commitment. The Science and Technology Ministry and CONICET, in the metropolitan center, have often been oblivious to the cultural and structural problems of the southern region. Ambitious programs have been publicized, but since they have little substance, remained real only on paper. There are valuable research centers in the southern region like CENPAT (Centro Nacional Patagónico) and CADIC (Centro Austral de Investigación Científica), but it has been difficult to replicate the kind of effort and scale of investment required.

The Climate Change Turn

The installation of the global climate change agenda has meant that local controversies and conflicts are now perceived as matters of global scope: the impacts of extractivism not only affect local communities but also generate global chains of negative impacts on fragile local ecosystems. It is interesting that this process of decentering of research lines related to climate change begins to articulate a micro-macro-micro logic. This logic concerns relationships between levels, which enhance the need to understand the conditions and/or impacts on local ecosystems and their relationship with anthropic phenomena. Thus, a process of decentering is activated that requires transdisciplinary views. Scientific agents are articulated to build these agendas from their own institutions, and in turn they begin to incorporate the issue of climate change into their policies and instruments. In the case of Aysén and Magallanes, these are research lines on marine and terrestrial ecosystems, sustainability, climate change, and their progressive link to R+D+I (research, development, and innovation) programs more linked to production issues: the generation of environmentally friendly products and prototypes, of networks of local organizations for the defense of the environment, and of awareness of the local and global conditions and effects of climate change. Thus, science, technology, and innovation (STI) policy priorities and agendas begin to be built that decentralize national priorities, localizing and simultaneously globalizing scientific concerns.

In Argentine Patagonia, research centers and universities work on issues based on the criteria of conservation of the natural environment and of minimizing negative impacts in terms of pollution and loss of biodiversity, often done in interdisciplinary collaborative studies (Kreps, Pastur, and Peri 2012). The national government has taken a stand, joining the global effort by implementing policies and strategies to reverse the greenhouse effect. These policies are aimed at the installation of large wind farms, an action that has been going on for several years through private and public investments, activated by incentive policies.

The STI Policy Turn

Policies for STI have faced the tension between defining a set of specific priorities (usually done in a centralized way) of short term and scope, and/or the construction of agendas or R+D+I programs sustained over time. On the other hand, scientific systems in Latin America tend to institutionalize the researcher's career (the Argentine case) or to encourage laissez-faire between scientific individuals who allocate resources and networks to support lines of research (Chilean case; Cancino et al. 2014). This generates two mechanisms that affect the dynamics of both the decentralization of research and of scientific agendas. The definition of local priorities (for example, around climate change issues) articulates networks of researchers for scientific collaboration, in such a way that they generate agreements around the positioning of their own research lines. Thus, cross-fertilization mechanisms between lines are generated and strengthened, inducing decentering. In this process, STI agendas or programs of greater local and temporal scope are built, activating forms of scientific decentralization. Thus, in Chile, the decentralization of agendas in the new setting seems to find a favorable scenario for its activation and growth as an organizing mechanism of scientific work in southern science.

In Argentina, since the second decade of this century, a different process can be perceived in terms of funding mechanisms from the national government. In various ways, STI is promoted with the aim of strengthening the academic system (Loray and Piñero 2014). Policy instruments favored the national reconstruction of the academic infrastructure and, in parallel, of the productive sector, especially depressed small and medium enterprises, in connection with tax regulations, training, and links with academia. They also aimed at strengthening territorial links by decentralizing the decision-making process, and also by allowing provincial and municipal governments to establish their local and regional agendas and accelerate cooperation processes with academic and social institutions in the territory. The dynamization of decision-making in these cases was conceived as linked to the construction of a fabric of intraregional collaboration, supposed to allow decentralizing research agendas through the actions and capacities of decentralized actors in the local STI centers (Ministerio de Ciencia, Tecnología e Innovación Productiva 2020).

Conclusion: From Decentering to Decentralizing Territorialities and Natural Research Agendas

A very relevant dimension emerges from the analysis carried out. It is about the existence of sociotechnical conflicts or controversies that condense the tensions between the development model, the scientific agendas, and the natural specificities of territories crossed by global and local concerns.

They have implications for political decision-making and the construction of national policies that allow reconciling multiple and often conflicting agendas. Those tensions also play a role of articulator and mirror for the activation of mechanisms that make the modalities and forms of agenda production move from logics to processes of decentralization. First, they set the public and scientific agenda, then they become an argument for building local scientific agendas, as in the documented history of socio-environmental or sociotechnical conflicts (Broitmann and Kreimer 2018; Fornillo and Nuñez 2021), the emergence of energy communities within the framework of energy transitions (Baigorrotegui 2018), and the problems of environmental pollution and resistance to technological change (Boso et al. 2020). We have seen how the case for or against salmon farming is politically framed and negotiated in contested cases. The prohibition of salmon farming in Tierra del Fuego, Argentina, was justified on scientific grounds (García et al. 2020; Macchi and Vigliano 2014). Local researchers reported on negative impacts on the environment, an issue that enhanced the scientific position and hence the political institutionalization of a decision. In this way, the scientific position on the negative impact of the activity reinforced a process of local political decentralization. The *ex ante* positioning of the scientific voice expressed concern at the expected impacts and effects of salmon farming. Subsequently, the economic impact of salmon farming versus the environmental impact in the territory became a controversy (Sanguinetti et al. 2014). In Chile, by contrast, looking for science-based solutions to the negative effects of activity resulting from intense industrial activity, there is a more frequent intervention of scientific action in the form of a specialized voice. Thus, a split results between what can be identified as more naturalistic and more applied research orientations showing the variety of purpose of the local scientific world in the attempt to establish authoritative knowledge and thereby which knowledge systems and corresponding procedures should be considered valid.

The ongoing political debates in Chile and Argentina are a major part of the concern about the spatial organization and use of the territory in both countries. Many stakeholders believe that coexistence is impossible between salmon culture, tourism, mining, and logging in the same territory, putting the environment and human welfare at risk. In short, all these cases are part of larger national debates related to the expansion of the exploitation frontiers and, consequently, to probable social, environmental, and climate implications.

We have shown ways in which decentralization has strong socioterritorial roots and is activated around a naturalistic and responsible imprint in science, in such a way that it can generate decentralized knowledge thanks to the operation of social, political, and global turns around the problem of climate change. To realize this, material and symbolic processes that ultimately aim to rebalance epistemic power relations must be created, stabilized, and extended. Epistemic decentralization seeks to complexify heterogeneous actors, knowledge, and policies through its expansion, densification, and rearticulation, a process in which decentered local scientific agendas emerge. In short, decentralization aims to give rise to broader, more interconnected networks—especially at the margins—in which power has been redistributed. Decentered knowledge seems better able to face today's great challenges. Distributed agency, produced through decentralized decision-making processes, may give rise to different types of agency (emancipatory, ethical, etc.) responding to the need of recognizing other subjects as knowing subjects within epistemic communities. As ideas embodied in texts, agendas are decentered through decentralized actors and capacities.

Notes

Dr. Ronald Cancino thanks FONDECYT Regular 1220219, from ANID-Chile for the project The Epistemological, Historical, and Territorial Construction of the Southern Zone as a Natural Laboratory: Scientific Agendas, Knowledge Networks, and Global Imaginaries, of which this publication is part.

1   As a result of the social outrising (the movement and lived citizen protest that began on October 11, 2019), a new constitution is being discussed today in Chile. In this process, the Commission for Knowledge Systems, Science and Technology, Culture, Art, and Heritage was created, to define an institutionality, policies and forms of financing, the role of the state, and the rights to culture, knowledge, and applications. Unfortunately, these proposals were not approved by the citizens. For more details, see Comisiones Convención Constitucional, https://www.cconstituyente.cl /comisiones/comision_integrantes.aspx?prmID=31.

2   For this section we rely mostly on Bilbao et al. 2020.

3   European experts recommended reducing the vulnerability of the plantations by managing the forest mass—that is, by promoting a mosaic with different species, ages, and densities—while simultaneously advancing the opening up and maintenance of the network of prevention infrastructures defined in plantation development plans (González et al. 2011, 2020).

4　See Law 26.815 of the Plan Nacional de Manejo del Fuego, which regulates the responsible organizations in charge of fire prevention and attack
in the natural environment (Ministerio de Justicia de la Nación 2013).

5　Several laws were sanctioned in the provinces of Neuquén, Río Negro,
Chubut, Santa Cruz, and Tierra del Fuego in order to comply with the
National Law of Fire Management. Intentional fires were main causes of
the loss of natural forests. Sometimes, under the Law of Natural Protected Areas in National Parks and peri-urban cities, also astonishing
landscapes used as recreational parks were included.

6　By Patagonian science, or *ciencia Patagónica*, we refer to the existing scientific capacities in the extreme south of Chile and Argentina. The main
characteristic of this science is its orientation toward the understanding
of natural, socioenvironmental, and biocultural phenomena typical of the
Patagonian territory. In both countries, there are similar identities and
lines of research, including strong relationships of scientific collaboration
between the Argentine and Chilean communities.

See, for example, the Autoridad Interjurisdiccional de las Cuencas
(AIC), established through institutional agreement among the governors
of the provinces of Neuquén, Buenos Aires, and Rio Negro; http://www
.aic.gov.ar/sitio/laaic.

References

AAICI. 2020. *Invertir en Argentina: Acuicultura.* Buenos Aires: Agencia Argentina de Inversiones y Comercio Internacional (AAICI), Ministerio de
Relaciones Exteriores, Comercio Internacional y Culto.

Baigorrotegui, Gloria. 2018. "Energy Communities in Patagonia: So Far So
Close from Extractivism." *Estudios Avanzados* 29: 56–74.

Bilbao, Bibiana, Lara Steil, Itziar R. Urbieta, Liana Anderson, Carlos Pinto,
Mauro E. González, Adriana Millán, et al. 2020. "Wildfires." In *Adapta-
tion to Climate Change Risks in Ibero-American Countries—RIOCCADAPT
Report*, edited by José Manuel Moreno, Clara Laguna Defior, Eduardo
Calvo Buendía, José Antonio Marengo, and Úrsula Oswald, 435–96.
Madrid: McGraw Hill.

Boso, Álex, Jaime Garrido, Boris Álvarez, Christian Oltra, Álvaro Hofflinger,
and Germán Gálvez. 2020. "Narratives of Resistance to Technological
Change: Drawing Lessons for Urban Energy Transitions in Southern
Chile." *Energy Research and Social Science* 65: 101473. https://doi.org.10
.1016/j.erss.2020.101473.

Broitman, Claudio, and Pablo Kreimer. 2018. "Knowledge Production, Mobilization and Standardization in Chile's HidroAysén Case." *Minerva* 56 (2):
209–29. https://doi.org/10.1007/s11024-017-9335-z.

Cancino, Ronald, Juan Carlos Aravena, Trace Gale, Eduardo Barros, Laura
Sánchez, Flavia Morello, et al. 2021. "Ciencia, territorio y sociedad en

Aysén y Magallanes: Informe diagnóstico del nodo ciencia austral a ANID-Chile." Chile: Universidad de Magallanes, Universidad de Aysén, INACH, CIEP.

Cancino, Ronald, Luis Antonio Orozco, Ricardo Bonilla, José Cóloma, and Fabian Cristian Ruiz. 2014. "Formas de organización de la colaboración científica en América Latina: Un análisis comparativo del sistema chileno de proyectos y el sistema colombiano de grupos de investigación." In *Perspectivas latinoamericanas en el estudio social de la ciencia, la tecnología y el conocimiento*, edited by Pablo Kreimer, Hebe Vessuri, Léa Velho, and Antonio Arellano, 380–95. Mexico City: Siglo XXI editores.

Carciofi, Ignacio, and Luciano Rossi. 2021. Acuicultura en Argentina: Red de actores, procesos de producción y espacios para el agregado de valor en búsqueda del impulso exportador para los productos acuícolas. Documento de Trabajo N° 13. Buenos Aires: Consejo para el Cambio Estructural, Ministerio de Desarrollo Productivo.

CONAF. 2017. Análisis de la afectación y severidad de los incendios forestales ocurridos en enero y febrero de 2017 sobre los usos de suelo y los ecosistemas naturales presentes entre las regiones de Coquimbo y Araucanía de Chile. Informe técnico. Santiago: Corporación Nacional Forestal (CONAF), Ministerio de Agricultura.

Connell, Raewyn. 2007. *Southern Theory: The Global Dynamics of Knowledge in Social Science*. Cambridge: Polity.

de Sousa Santos, Boaventura. 2018. *The End of the Cognitive Empire: The Coming of Age of the Epistemologies of the South*. Durham, NC: Duke University Press.

Fornillo, Bruno, and Jonatan Nuñez. 2021. "Aysén reserva de vida: Energía, mercantilización y resistencias en la Patagonia chilena." *Letras Verdes, Revista Latinoamericana de Estudios Socioambientales*, no. 29: 65–81.

Fundación Vida Silvestre Argentina. 2010. "Minería metalífera de mediana y gran escala en la Argentina: Posición de la Fundación Vida Silvestre Argentina." World Wildlife Foundation Argentina, March. https://wwfar.awsassets.panda.org/downloads/posicion_fvsa_mineria_final.pdf.

García, Juan Ignacio, Carolina Hernández, and Silvina A. Romano. 2020. "Análisis de la acuicultura de salmónidos intensiva de gran escala en el canal Beagle como estrategia para el desarrollo de Tierra del Fuego." *Estudios Económicos* 37 (74): 161–90.

Gobierno de Chile. 2017. *Plan de acción para la recuperación de patrimonio natural y productivo afectado por los incendios de 2017*. Santiago: Gobierno de Chile, Ministerios de Agricultura, de Hacienda, de Economía, Fomento y Turismo, del Ambiente. https://www.camara.cl/verDoc.aspx?prmID=103379&prmTIPO=DOCUMENTOCOMISION.

González, Mauro E., Antonio Lara, Rocío Urrutia, and Juvenal Bosnich. 2011. "Cambio climático y su impacto potencial en la ocurrencia de incendios forestales en la zona centro-sur de Chile (33°–42°S)." *Bosque* 32 (3): 215–19.

González, Mauro E., R. Sapiains, S. Gómez-González, R. Garreaud, A. Miranda, M. Galleguillos, M. Jacques, et al. 2020. *Incendios forestales en Chile: Causas, impactos y resiliencia.* Santiago: Centro de Ciencia del Clima y la Resiliencia (CR)2, Universidad de Chile, Universidad de Concepción y Universidad Austral de Chile. http://www.cr2.cl/wp-content /uploads/2020/01/Informe-CR2-IncendiosforestalesenChile.pdf.

Herrera, Marco. 2021. "Controversias: Nuevas demandas para la CTCi en Aysén y Magallanes; Informe al Nodo Ciencia Austral." Unpublished ms.

Kreps, Gastón, Guillermo Martínez Pastur, and Pablo Luis Peri. 2012. *Cambio climático en Patagonia Sur: Escenarios futuros en el manejo de los recursos naturales.* Buenos Aires: Instituto Nacional de Tecnología Agropecuaria.

Law, John. 2007. "Actor Network Theory and Semiotics." Heterogeneities.net: John Law's Web Page, April 25. http://heterogeneities.net /publications/Law2007ANTandMaterialSemiotics.pdf.

Loray, Romina, and Julio Fernando Piñero. 2014. "El Plan Argentina Innovadora 2020: Avances en materia conceptual e institucional de las políticas públicas en ciencia, tecnología e innovación (CTI) de la Argentina reciente." *Jornadas de Sociología de la UNLP* (Universidad Nacional de La Plata) 8: n.p.

Macchi, Patricio J., and Pablo H. Vigliano. 2014. "Salmonid Introduction in Patagonia: The Ghost of Past, Present and Future Management." *Ecología Austral* 24 (2): 133–264. https://doi.org/10.25260/EA.14.24.2.0.19.

Machado, Horacio, Maristella Svampa, Enrique Viale, Marcelo Giraud, Lucrecia Wagner, Mirta Antonelli, Norma Giarracca, and Miguel Teubal. 2011. *15 Mitos y realidades de la minería transnacional en Argentina: Guía para desmontar el imaginario prominero.* Buenos Aires: CLACSO.

Mineduc (Ministerio de Educación, Chile). 2023. *Informe 2023: Matrícula en Educación Superior.* https://www.mifuturo.cl/wp-content/uploads/2023 /07/Matricula-_en_Educacion_Superior_2023_SIES.pdf.

Ministerio de Ciencia, Tecnología e Innovación Productiva (Argentina). 2020. *Argentina Innovadora 2020: Plan Nacional de Ciencia, Tecnología e Innovación Lineamientos estratégicos 2012–2015.* https://www.argentina .gob.ar/sites/default/files/pai2020.pdf.

Ministerio de Economía, Fomento y Reconstrucción (Chile). 1992. "Decreto 430: Fija el texto refundido, coordinado y sistematizado de la ley n° 18.892, de 1989 y sus modificaciones, ley general de pesca y acuicultura." Biblioteca del Congreso Nacional de Chile: Ley Chile. https://www.bcn .cl/leychile/navegar?idNorma=13315.

Ministerio de Justicia de la Nación (Argentina). 2013. "Ley del Manejo del Fuego Nacional Nª 26.815." https://servicios.infoleg.gob.ar /infolegInternet/anexos/205000-209999/207401/texact.htm.

Ministerio de Minería (Chile). 1982. "Ley 18097: Ley Organica Constitucional sobre Concesiones Mineras." Biblioteca del Congreso Nacional de Chile: Ley Chile. https://www.bcn.cl/leychile/navegar?idNorma=29522.

Mouillot, Florent, and Christopher B. Field. 2005. "Fire History and the Global Carbon Budget: A 1° × 1° Fire History Reconstruction for the 20th Century." *Global Change Biology* 11 (3): 398–420.

Rafaele, Estela, Martín Núñez, and María Relva. 2015. "Plantaciones de coníferas exóticas en Patagonia: Los riesgos de plantar sin un manejo adecuado." *Ecología Austral* 25 (2): 86–157.

Román, Álvaro. 2020. "Integración territorial como marginación: Obstáculos para las zonas aisladas en Aysén y Magallanes, Chile." *Revista LIDER* 37 (22): 77–99.

Sanguinetti, Javier, Leonardo Buria, Laura Malmierca, Alejandro E. J. Valenzuela, Cecilia Núñez, Hernán Pastore, Luis Chauchard, et al. 2014. "Manejo de especies exóticas invasoras en Patagonia, Argentina: Priorización, logros y desafíos de integración entre ciencia y gestión identificados desde la Administración de Parques Nacionales." *Ecología Austral* 24: 183–92.

Schlichter, Tomás, and Pablo Laclau. 1998. "Ecotono estepa-bosque y plantaciones forestales en la Patagonia norte." *Ecología Austral* 8: 285–96.

Subsecretaría de Estudios y Prospectiva (Argentina). 2019. *Anuario estadístico de la República Argentina*, vol. 34. Buenos Aires: Ministerio de Ciencia, Tecnología e Innovación, Dirección Nacional de Información Científica.

Wagner, Lucrecia S. 2016. "Conflictos socioambientales por megaminería en Argentina: Apuntes para una reflexión en perspectiva histórica." *AREAS Revista Internacional de Ciencias Sociales* 35: 87–99.

A State-Led Strategy of Decentralization

The BRICS Experience

From various theoretical and methodological perspectives, older and more recent studies on the production and circulation of academic knowledge describe hierarchies and inequalities on a global scale. Until rather recently, such studies largely converged on the idea that Western Europe and North America were central in the academic worlds, and that what was perceived as their domination marginalized research in the Global South and in languages other than English.[1] However, the global academic landscape seems to be changing recently. Significant effort has been mobilized to create bridges between southern countries from the 1990s onward. While the fundamental problem of hierarchical and inequality structures persists, and the various terminologies suggested to understand them remain relevant, we are currently observing a shift from North Atlantic centrism toward multipolarity.[2]

Analysts have clearly observed the rise of Asian science, led by China (Kahn 2015, 106). China's gross expenditure on research and development (408.8 billion USD in 2015) was comparable to the total for the twenty-eight EU countries (386.5 billion USD) and approaching that of the United States (502.9 billion USD). In 2015, China was the country with the largest absolute number of research staff (1.619 million full-time equivalents, compared with 1.380 million in the United States and 1.841 million in the EU; Shashnov and Kotsemir 2018, 1125). But not only China has moved to the forefront of such international comparisons. Before the formation of the BRICS alliance, during the 1990–2010 period, "the governments of the BRICS countries boosted their investments in research and development to become part of the group of nations doing research at the highest level" (Bornmann, Wagner, and Leydesdorff 2015, 1511).

Today, the BRICS countries (Brazil, Russia, India, China, and South Africa) not only rise as single scientific nations but collectively lead and organize this transformation toward a more multipolar global science system through their transnational alliance. One perception from the BRICS perspective, for instance, claims that "acting on the world stage as a unified group, the BRICS countries are becoming one of the world centers of producers of new knowledge, modern technology, and innovative development . . . while the USA is losing their role of 'scientific superpower'" (Shashnov and Kotsemir 2018, 1116, 1127).

Some European observers feel upset, if not threatened: "Like a bulldozer, Chinese science overthrows the hierarchies that had been established in the last century. It imposes itself as a big science power and promises to become a superpower. The US loses its hegemony that had been blatant half a century ago. Japan collapses. New countries emerge: India, Iran, Brazil, South Korea. And France? France is now only on the 7th rank, outpaced by China, but also India, and represents no more than 3.2% of global scientific publications" (Huet 2018).[3]

Development of the BRICS countries goes along with their ascendance as emerging powers at the economic, political, and cultural levels. Since the meltdown of the global economy in 2007–8, BRICS has employed a rhetoric that calls for the creation of significant alternatives to the existing world order—more just, fair and equitable, multipolar, democratic and representative, South-South (Coning and Puri 2017, 92). Theoretical reflections have paralleled the BRICS effort to a new Bandung (Oustinoff 2017), or a new counterdependency (Muhr and Azevedo 2018, 524), or have interpreted it as de-Westernization (Mignolo 2012).

In order to enact a different world order, the BRICS group's cooperation has created multilateral institutions. The New Development Bank, founded in 2016, demonstrates that not only have the BRICS countries become important economic powers, but they also act on the global economy. They have shifted the balance of power in the UNESCO World Heritage Committee (Bertacchini, Liuzza, and Meskell 2015). With Chinese internet giants behind its gigantic virtual wall and India having become the second largest nation of internet users in 2016, BRICS is also considered to represent a challenge to US domination in media and communication (Thussu and Wirth 2017, 67). However, the project of materially rewiring the five nations with direct internet cables, to replace their indirect connection via the United States in order to render them numerically independent as a reaction to the Snowden affair, has failed (Zyw Melo 2017).

Within this multidimensional framework of cooperation, science and higher education are another important brick that we interpret here as an attempt at decentralization at the level of international geopolitics. The five governments have agreed on a long-term, state-driven strategy to intensify cooperation at the levels of higher education and scientific research. They have put in place major public policy initiatives under the responsibility of their ministries of education (Li 2018, 394) and ministries of science and innovation. The following analyzes these two different science policy fields—education (excluding basic education) and science, technology, and innovation (STI; excluding innovation)—under the assumption that they represent a determined strategy at decentralization.

An Example of Decentralization

The editors of this volume suggest *decentralization* as a counterweight to *decentering*. While decentering focuses on theoretical and epistemological issues, decentralization refers to the "hard currencies," resources and institutions. The calls for decentering knowledge against historically grown Eurocentrism have transformed, partly, the higher education and research practices and institutions across Europe and North America but have not provoked any substantial changes regarding global hierarchies and divides in global knowledge production and circulation. Ari Sitas made this clear in a 2006 text: "Unfortunately, the emphasis on discourses (and texts), their constructions and inventions encouraged by postcolonial theorists, despite their critical and emancipatory promise, prove to be frustrating. By profiguring processes of signification and discursive power, they leave the 'steering media' of money and power and more importantly the institutional matrices that constrain social life and indeed their own claims, untouched" (2006, 362). Decentralization shifts attention toward precisely those "steering media." As suggested by the editors' introduction to this volume, it implies incorporating actors who had been marginalized, ignored, underestimated, or directly eliminated into knowledge-producing systems. This leads to the complexification, enlargement, and augmenting of existing assemblages. The editors highlight, as one of the reasons why peripheral knowledge remains underrepresented in the centers of global academia, the lack of infrastructures required to enable such processes of integration, that is, there is a lack of material and symbolic resources to participate in peer-to-peer dialogues on an equal footing.

In our study of BRICS, we consider decentralization to be first and foremost a political program that aims at the creation, stabilization, and extension of material and symbolic processes that ultimately aim to rebalance asymmetrical epistemic power relations on a global scale. We believe that the BRICS alliance's research and higher education policies provide a good empirical example of concrete programs to achieve this. This way, we take up the challenge of this edited volume, in analyzing decentralizing powers through their actions and strategies, the possibilities and limitations that they put in place from the peripheries in order to increase their epistemic authority and recognition. In the following, we propose to take an inventory of the selected key initiatives that BRICS has put forth so far to achieve this, based on an analysis of official documents.

Empirical Approach

This chapter proposes a systematic analysis of policy documents. The ongoing pandemic context has motivated our choice to rely on official documentation that is accessible online, since other approaches that we had initially envisaged for this project (an international workshop, followed by face-to-face qualitative interviews and group discussions, as well as field trips for qualitative interviewing) were not feasible or have been considerably delayed. The analyzed documents were retrieved from the website of the BRICS Network University (http://nu-brics.ru; now https://mspo .hse.ru/en/nubrics/documentbrics)[4] and that of the BRICS STI initiative (http://brics-sti.org/) in November 2021.[5] Significant parts of the documentation on education refer to general, vocational, and technical education and are therefore left out of the following analysis that focuses on higher education only. The very complete online documentation of the BRICS Information Centre at the University of Toronto (http://www.brics .utoronto.ca/) allowed us to fill in the gaps of some missing documents.

The limitations of a document-based analysis are obvious: policy documents are the final, official results of complex negotiations between diverse actors who pursue various and oftentimes adverse strategies in a field of power. The analyzed documents were released with the agreement of the highest national authorities in the field in each of the five countries. The global decentralization process that we observe corresponds to a centralized process at the national levels. However, if our argument is about global decentralization from a geopolitical perspective, focusing on a state-to-state alliance as the key actor, then we do believe that taking this official voice seriously does make sense. We will also tentatively draw on first insights from the ongoing qualitative interviews regarding the interpretation of results.

Toward the BRICS Network University: Document Analysis

Higher education was placed on the BRICS agenda when the ministers of education met for the first time in Paris in 2013, in the context of the thirty-seventh session of UNESCO's General Conference. The purpose of the meeting was "to identify potential areas of education collaboration between the BRICS countries, and agree on implementation mechanisms" ("Minutes of BRICS Education Ministers and UNESCO Meetings" 2013). The 2017 Beijing Declaration augmented the initial vision, declaring higher education collaboration as being of significance for the overall BRICS partnership and in particular for "people to people exchanges" (BRICS Ministers of Education 2017, 1). Gradually, higher education was embedded in the broader agenda of the BRICS alliance (UNESCO 2014, 20).

The BRICS Higher Education Agenda

The first meeting of ministers of education foreshadowed four key issues. A closer look at the documentation reveals that, initially, the ministers of the five countries put very different emphases on those four points ("Minutes of BRICS Education Ministers and UNESCO Meetings" 2013, quoted in the list that follows). Analysis of the subsequent meetings also shows that the initial four suggestions were not followed up to the same degree:

1   The encouragement of academic exchange in the form of mobility was put on the agenda by Yuan Guiren (China), in relation to the question of mutual recognition of qualifications as a precondition for student mobility between BRICS countries, and by Aloizio Mercadante (Brazil), who suggested exchange of academics between BRICS countries.

2   Academic collaboration was favored by the suggestion to develop more "institutional collaboration" (Brazil), or to strengthen co-

operation in and through the BRICS Academic Forum and BRICS Think Tanks (Minister Angie Motshekga, South Africa). China was singular in calling for "joint research and copublishing of scientific results by BRICS academics and universities." It was Brazil that put the "establishment of a network of BRICS elite universities" on the agenda. This proposition was the starting point for the Network University (NU).

3   The issue of developing comparisons between educational systems was shared by India, South Africa, China, and Russia. Minister Shashi Tharoor (India) suggested a single point to the ministers' meeting, namely "exchanges about respective assessment systems and analyses of outcomes," a rather open-ended proposition. China, in turn, called for "development of common tools to ensure the quality of education within BRICS" and South Africa for "comparative research and information on instruments for the measurement and monitoring of the quality of education," which put the question of quality assessment in education on the table. Minister Dmitry Livanov (Russia) insisted on this single point of comparison between countries and framed it in a particular way: "development of a ranking system for BRICS universities in order to enhance their international status" and "comparative studies on the quality of education . . . within the BRICS countries to enable benchmarking." Clearly, the Russian agenda, at this stage, focused only on the questions of international ranking and benchmarking—that were absent for Brazil.

The 2015 Brasília declaration by ministers of education stressed "the paramount importance of the development of joint methodologies for education indicators to support decision making in BRICS member states," recognizing "that the indicators . . . should be based primarily *on national assessments, administrative data and national household surveys instead of extension of existing international surveys*" (BRICS Education Ministers 2015, emphasis added). Benchmarking and ranking were abandoned in favor of "the development of common principles of accreditation and quality assurance," but we do not get any further insights on what those common principles or indicators would encompass and how they should be applied. These passages clearly indicate how the BRICS countries were struggling to find a way between two opposing goals: competition in the global race for excellence as opposed to building a real alternative within the global higher education arena.[6]

4 South Africa's idea of joint development of an international strategy in the domain of education was not taken up in any of the subsequent documents, apart from closeness of the BRICS agenda to the UNESCO programs for education.

Out of these divergent positions on an agenda for BRICS cooperation in the domain of education, only three points were retained as recommendations by the first ministers' meeting: mutual recognition of qualifications, exchange of academics and students, and the promotion of the development of networks of BRICS universities ("Minutes of BRICS Education Ministers and UNESCO Meetings" 2013). The following focuses on the project of the Network University, as it represents the institutionalization of academic collaboration and mobility.[7]

Creation of the Network University

The 2015 Brasília declaration created a working group to elaborate the modalities of a BRICS Network University. Eight months later in Moscow, ministers signed a memorandum of understanding (MoU) preparing for the institutionalization of intra-BRICS academic mobility, exchange, and joint research. It defined the Network University as "an educational project aimed at developing, preferentially, bilateral/multilateral short-term joint training, master's and PhD programs along with joint research projects in various knowledge fields according *to common standards and quality criteria,* given recognition of the learning outcomes by BRICS NU participants *as per national criteria*" (BRICS Ministries of Education 2015, 2, emphasis added). The formulation highlights the fundamental BRICS principles of joint effort while maintaining national autonomy and sovereignty. Accordingly, the NU participants are national higher education institutions that maintain their full autonomy and that interact on an equal footing, mutually recognizing national regulations and practices (BRICS Ministries of Education 2015, 2).

Interestingly, one of the key aims was the education of highly qualified professionals, "who are capable of combining traditional knowledge with science and contemporary technologies"—an idea that was not taken up again in subsequent documents. At the foundation stage, each ministry could nominate a maximum of twelve institutions as participants of the BRICS NU. This corresponds to a strongly centralized process at the national levels of the five countries.

The same applied to the structures of governance. The NU is regulated by an international governing board composed of representatives of education ministries and of NU participants from each BRICS national coordination committee, and by those same national coordination committees, created by ministries of education to ensure the country-wide management of the BRICS NU. The national coordination committees are composed of NU participants and two representatives of ministries of education. They may also include experts and representatives of the business community, civil society, or international organizations. To sum up, the NU "is not a supranational institution but an international structure among BRICS NU National Coordination Committees, created in each member state by the education ministries, and financed by the participating universities" (Muhr and Azevedo 2018, 530).

Finally, the NU consists of international thematic groups according to "knowledge field priorities," closely related to BRICS cooperation in science, technology, and innovation (see next section). They include energy, computer science and information security, BRICS studies, ecology and climate change, water resources and pollution treatment, and economics. Article 7 of the MoU relates to the creation of curriculum: "The BRICS NU participants will work out the details of: structure and content of educational programmes; mutual recognition of the training outcomes; academic mobility forms as decided by the International Thematic Groups on the knowledge field priorities of the BRICS NU; procedures for admission; principles of educational process arrangement; issues of interim and final certification which are regulated by agreements between BRICS NU participants on joint training of highly qualified personnel" (BRICS Ministries of Education 2015, 4). Beyond these rather technical points, we do not learn more about either the contents or practical procedures to determine these.

Article 13 addresses the issue of funding. Again, according to the principle of national autonomy, the NU's activities within each country should be financially covered by the member universities' own financial means, that is, by each country independently. An institutional website, hosted by Russia (nu-brics.ru), was created.

In 2016, ministers of education decided that an annual conference of the NU should be convened in the home country of the BRICS Chair (BRICS Ministers of Education 2016). The first annual NU conference in

2017, under the title "Pragmatic Cooperation and International Education," declared: "We believe that participating in the BRICS NU will play an important role in enhancing cooperation and exchanges among all member universities, strengthening their *visibility, impact* and *competitiveness*" (BRICS Network University 2017, emphasis added).

BRICS STI Collaboration: Document Analysis

In order to assess the science, technology, and innovation (STI) collaboration between BRICS, it is necessary to look at the strategic level, that is, the ministerial meetings, starting in 2014; at the MoU on STI collaboration, signed in 2015; and at the two key actions to enhance scientific research collaboration, namely the joint calls within the BRICS STI Framework Program, as well as the activities carried out by the existing thematic working groups. A last section deals with two further initiatives in the domain.[8]

Ministerial Meetings

The ministers of STI met for the first time in Cape Town in 2014 "to discuss and coordinate positions of mutual interest and identify directions of *institutionalizing* cooperation in science, technology and innovation within the framework of BRICS" (I BRICS Science, Technology and Innovation Ministerial Meeting 2014, emphasis added). On the one hand, ministers wanted STI cooperation to be "gradual" and "pragmatic," and "the significance of competitiveness" was acknowledged, in line with mainstream assumptions on the economic usefulness of science in the global capitalist market. On the other hand, it should reflect "the principles of openness, solidarity and mutual assistance" and be "people-centred" and "public-good driven" (I BRICS Science, Technology and Innovation Ministerial Meeting 2014), expressing a commitment to an alternative agenda.

The second meeting, held in Brazil in 2015, clearly prioritized the aim of economic competitiveness in the global arena while maintaining the idea of economic complementarity within BRICS. Achieving competitiveness was based on a rather conventional logic of catching up, phrased as "bridg[ing] the scientific and technological gap between BRICS and developed economies" (II BRICS Science, Technology and Innovation Ministerial Meeting 2015). The following actions were to achieve these goals: "cooperation in the framework of major research infrastructures; coordination of existing large-scale national programmes of BRICS countries; setting up a Framework Programme for funding multilateral joint projects for research, technology commercialization and innovation; establishment of a joint Research and Innovation Networking Platform" (BRICS Russia 2020, 8). In July 2015, the national funding agencies of the five countries met to discuss the creation of a multilateral program to fund joint research projects (Sorokotyaga 2021). The third ministerial meeting, held in Moscow in 2015, established the BRICS STI Framework Programme. The BRICS STI Funding Working Group should ensure regular exchange between national funding agencies in order to coordinate it.

Ministers have met on a yearly basis since then. The seventh meeting highlighted the importance of the BRICS STI initiative for global knowledge production: "The increasing interaction between our researchers, academies and laboratories can bring an important new perspective and contribution to the world scientific production" (VII BRICS Science, Technology and Innovation Ministerial Meeting 2019).

At the same time, one perceives between the lines certain critical points of the initiative. On the one hand, the declaration openly demanded more funding and simpler procedures if long-term sustainability at the level of STI cooperation was to be achieved. This demand directly addressed the governments of the five countries. The declaration also reads like a rejection of too much direct influence of government officials in collaboration between scientists.[9] The 2019 Campinas meeting outlined with more clarity the various levels of the overall structures of the STI initiative, composed of ministerial and senior officials meetings; thirteen thematic working groups (see below); joint calls for research projects (see below); as well as other initiatives, such as the BRICS Young Scientists Forum, Water Forum, BRICS Science Academies Meeting, Conference on Technology Foresight and STI Policy, the platform for research infrastructure collaboration, and the Action Plan for Innovation Cooperation (BRICS Russia 2020, 12).

Memorandum of Understanding on STI Collaboration

The MoU on collaboration between member states in the STI domain, signed in Brazil in 2015, enacted the BRICS STI initiative. As a strategic framework for cooperation, it laid the grounds for its development and institutionalization in the subsequent years. It based STI cooperation on the guiding principles of "voluntary participation, equality, mutual benefit, reciprocity and subject to the availability of earmarked resources for collaboration by each country" (MoU on STI Cooperation 2015).

Article 4 of the MoU outlined the mechanisms and modalities for the planned cooperation:

> (a) Short-term exchange of scientists, researchers, technical experts and scholars; (b) Dedicated training programmes to support human capital development in science, technology and innovation; (c) Organization of science, technology and innovation workshops, seminars and conferences in areas of mutual interest; (d) Exchange of science, technology and innovation information; (e) Formulation and implementation of collaborative research and development programmes and projects; (f) Establishment of joint funding mechanisms to support BRICS research programmes and large-scale research infrastructure projects; (g) Facilitated access to science and technology infrastructure among BRICS member countries; (h) Announcement of simultaneous calls for proposals in BRICS member countries; (i) Cooperation of national science and engineering academies and research agencies. (MoU on STI Cooperation 2015)

All of those have been realized, to varying degrees and with different emphases. The following two sections outline the key instruments to enhance scientific cooperation through a collective funding mechanism and through the establishment of transnational thematic working groups that define funding priorities.

Research Collaboration: Thematic Working Groups

The STI thematic working groups are one key initiative of STI cooperation directly related to the joint calls for projects (see below). During their first meeting, ministers had defined five thematic priority areas and respective leaderships by one of the five countries: climate change and natural disaster mitigation (Brazil); water resources and pollution treatment (Russia); geospatial technology (India); new and renewable energy, and energy efficiency (China); and astronomy (South Africa). By 2019, nine working groups existed, since Biotechnology and Biomedicine, Material Sciences and Nanotechnology, Information and Communication Technologies and High Performance Computing, Ocean and Polar Science and Technology, and Photonics had been added, but Water Resources and Pollution Treatment was discontinued. Their task was more clearly defined: "In the thematic working groups, officials and researchers define lines of action;

develop joint projects; exchange knowledge and experience in their areas of study; and decide the scope of the joint calls" (VII BRICS Science, Technology and Innovation Ministerial Meeting 2019). Within the STI framework, a close collaboration between the research community and governments is thus encouraged. In particular, scientists directly contribute to the elaboration of the joint calls when they formulate, as delegates of working groups, the thematic directions and scope of those calls. Cooperation in BRICS STI enables a particularly close, multilateral collaboration between key agencies in the domain of scientific and technological development: governments, scientists, and national funding agencies. Stakeholders from industry are associated in the area of innovation.

Research Collaboration: Joint Calls

Article 6 of the MoU, dedicated to the question of funding, established that STI cooperation was to be enabled through "appropriate BRICS country funding mechanisms, instruments and national rules," thus upholding the principle of national sovereignty. The BRICS STI Funding Working Group coordinates collaboration between nine national funding agencies.[10] In 2016, a pilot call was launched, encouraging researchers from the five countries to link up with colleagues in at least two other BRICS member countries in order to submit a joint research proposal in the thematic areas of the call. The first three calls received a total of 1,100 project proposals by more than 3,400 researchers, out of which ninety-one in total were selected for funding. In 2020, the Framework Programme launched a special call dedicated to interdisciplinary research on the multilevel impact of the COVID-19 pandemic, gathering 111 multilateral projects in response (Sorokotyaga 2021). For the first time, the social sciences were included for potential funding through the Framework Programme. According to the BRICS STI website, twelve joint projects out of a total of 111 submitted proposals were selected for funding (BRICS STI Framework Programme 2021). A fifth call was open until October 2021.

The website of the BRICS STI initiative has published short descriptions of all projects funded under the 2016 pilot call and the 2017 second call (BRICS STI Framework Programme n.d.). A total of fifty-eight projects have been funded, each with partners from at least three out of the five countries. In order to get an overall impression of the research funded within the STI program, the short project descriptions (text only, excluding technical information contained in tables) were fed into WordArt in

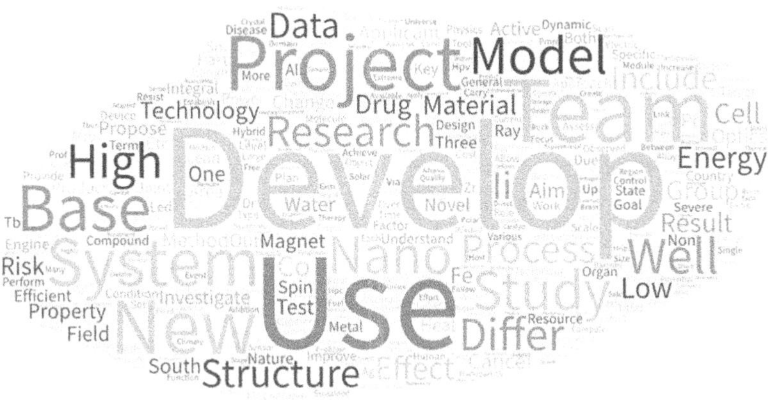

Figure 10.1 Word cloud based on project descriptions of collaborative research projects funded within the 2016 and 2017 STI framework programs. Source: the authors.

order to produce a word cloud. The country names of the five BRICS countries, obviously among the most frequent keywords in the texts, were deleted from the word frequency list in order to obtain a better image of the scientific content of the projects. The result is presented in figure 10.1.

The word cloud puts two terms center stage that seem to characterize the orientation of the funded projects: *develop* and *use*; that is, the application orientation is easily visible. The word cloud also shows the strong focus on hard sciences and, in particular, technology development. Some of the thematic priority fields clearly appear, such as water resources or energy, material sciences and nanotechnology, disaster mitigation (*risk*) or biomedicine (*drug*).

Joint Use of Research Infrastructure

The MoU had also foreseen cooperation at the level of research infrastructure, in particular in investment-intensive big science. A working group on megascience projects was set up. At its kickoff meeting hosted by Russia in May 2017, this working group launched an initiative called BRICS Global Research Advanced Infrastructure Network (GRAIN). As a first step, GRAIN started a joint platform hosted by the Joint Institute for Nuclear Research (Russia) as a digital opportunity for exchange of relevant information, among others on opportunities open to researchers from the BRICS domain in the form of partnerships or access calls for upcoming events (BRICS GRAIN: Research Infrastructure Platform; https://brics-grain.org/, accessed November 11, 2021). The collective use and development of big science projects formed part of the realization of BRICS STI cooperation.

Currently, GRAIN provides information on thirty large-scale infrastructures across the five countries: twenty-one operational, seven developing, and two planned. They include, among others, reactors, accelerators, electron colliders, and detectors in Russian nuclear physics; Chinese electron and positron colliders, reactors, and heavy ion research facilities as well as devices in China's laser technology; several Brazilian laboratories in the domain of research in energy and materials; large-scale telescopes in South Africa and India; and Indian radio and solar observatories.

According to the "STI Overview," the cooperation on megascience infrastructures holds significance beyond BRICS: "The platform . . . is intended to raise the awareness of the global community to the mega-science projects, implemented within the framework of BRICS STI Cooperation on the research infrastructures of BRICS countries by joint forces, fostered and developed by scientists and researchers from Brazil, Russia, India, China and South Africa" (BRICS Russia 2020, 35). The BRICS initiative is supposed to "engage the global research community to the BRICS Research Infrastructures" (35).

Results: Material and Symbolic Decentralization

The document analysis confirms, in our view, that the five countries have engaged in a political program toward decentralization. Of course, the studied documents express only the intentions of the BRICS authorities. From there to the realization of their strategy is still another step. The BRICS Vaccine C&D Center, for instance, was recommended for several years, was virtually launched in 2022, but as of 2024 has not materialized. Regarding the Network University, several of the involved actors have written or talked to us about their experiences. Their accounts allow us to draw a more differentiated picture of the challenges involved in putting this major project into place. We learn, for instance, about difficulties in constituting national delegations of equal sizes, and the impression left on some delegates of the 2016 Yekaterinburg meeting, that the Russian delegation had priorities different from those of other countries, namely in the first place the "promotion of international student mobility of their own universities," rather than cooperation between researchers of joint courses. We learn that the Chinese delegation was divided between proponents of the NU and those of the University League during the second meeting. While under the presidency of Dilma Rousseff, who held BRICS high on her agenda, Brazil had strongly pushed for the establishment of the NU, after her impeachment the new Ministry of Education turned hostile to the BRICS higher education project and reduced Brazilian participation. Financial questions had remained unresolved for a long time (Dwyer 2017, 103).

Moreover, the rapid pace of the development of the BRICS alliance in general did not leave enough time for reflection and for learning from failures and mistakes. According to participants in the national delegations, none of them had been properly prepared for the complexities of multilateral negotiations at that level. The main reason given for such failures, interestingly, is that "which diplomacy puts so much energy into hiding: the mechanisms of reciprocal disqualification and the strengthening of one's own identity" (Dwyer 2017, 105). Or, "in other words—it was quicker to fall into easy typifications that reinforce one's identity and 'conventionally adopted approaches in one's homeland' than to explore the others' frontiers and listen to the other . . . which is necessary in order to produce a deep understanding" (Tom Dwyer, pers. comm., November 2021).

The political strategy seems to favor, we argue, "decentralization," rather than "delinking" or "counter-hegemony" (Keim 2011). It has not generated, as far as the established priority fields are concerned, any revolutionary knowledge in the sense of fundamental epistemic innovations. Rather than anything that would resemble BRICS knowledge, the joint research pursued by BRICS actors does not seem to be different from conventional science and technology knowledge. Nuclear physics, ocean science, high-performance computing, telescopes, and so on, produce mainstream science and R&D, with the exception, eventually, of the social sciences in the domain of BRICS studies. The earliest document from the STI ministers' meeting mentioned "traditional knowledge" and the need to integrate it into STI, but this does not seem to have left any traces. What is new is the fact that this knowledge is produced on the basis of different, South-South patterns of collaboration that connect different scientific actors. Our analysis seems to confirm a critical observation made by Leandro Rodriguez Medina and Sandra Harding in their introduction to this volume: "Epistemic decentralizing has what we might call an indirect link to ideas. When more actors can produce and diffuse knowledge, it cannot be determined in advance that those ideas will contribute to the decentering of thought."

Our results are divided into several subsections: (1) decentralization as public policy; (2) differentiating between material and symbolic decentralization; (3) difficulties of a strongly state-led agenda; and (4) a qualification regarding the specificities of mega-science.

Material Decentralization as Public Policy

The analysis of BRICS documents shows that the national governments of the five countries pursue an explicit public policy toward increased cooperation in science and higher education. It has been the role of their governments, through their respective science and education ministries, to push for collaboration. Analysts have highlighted that "collaboration efforts among the BRICS countries may be more influenced by government-to-government efforts than mediated by markets" (Varghese 2015, 46). Because historically, cooperation has been stronger between BRICS and "developed countries," "government initiatives and public action are needed at this stage to promote cooperation and expand collaboration in higher education among BRICS countries" (46). The analyzed documents also express a heightened consciousness that the BRICS initiatives should change the global academic landscape, as the highlighted quotes demonstrate.

One of the key issues with decentralization is that those actors at the margins often do not have the necessary material or symbolic resources to participate in peer-to-peer dialogues on an equal footing. It seems to us that the BRICS example shows two things. Our analysis confirms that important material resources are made available, and there is the political willingness to invest in an academic decentralization effort. What has been achieved, or is about to be achieved, is a substantive institutionalization of decentralized higher education and research infrastructures with the potential to create alternative transnational flows of competency, knowledge, and staff. Indeed, the foundation of a transnationally networked institution without a dedicated physical space (the BRICS Center at Ural Federal University in Yekaterinburg has become its permanent secretariat, however), and without a budget on its own, in which national universities maintain their full autonomy, not least financially, represents a unique alternative institutional construction. One would be tempted to describe the NU as a dematerialized project. However, the difficulties in getting online teaching operational, given that Google does not work in China, for example, reminds us that supposedly virtual networks still rely on material infrastructures. The difficulties related to e-learning were only resolved as a consequence of the pandemic context. Such a functioning requires a "permanent dialogue and mobility between universities, professors and students of the five countries" (Dwyer 2017, 102). Whether the strategy succeeds or not, ultimately, does not seem to be a matter of resources. What is at stake, however, is the symbolic resources. Decentralizing those appears to be a different issue and depends on more than material resources.

The Challenge of Symbolic Decentralization

Some analysts remain hesitant about how to categorize the newly emerging powers. They represent, on the one hand, "growing and increasingly powerful academic systems" but at the same time "remain gigantic peripheries" (Altbach 2012, 128). Such quotes address the difference between material and symbolic decentralization. While the BRICS have engaged in the creation of alternative infrastructures that link formerly peripheral countries and have the potential to fundamentally transform the existing global academic landscape, organizing its multipolarity, they seem to find it more difficult to resolve symbolic decentralization. We mean by this that the historically built prestige of established centers of knowledge production remains firmly established. Decentralizing academic recognition is not only a matter of resources but requires a reconfiguration of visibility and reciprocal acknowledgment (see below).

While a strong level of academic and scientific development characterizes all five countries and turns them into regional centers, those science and higher education systems remain oriented toward the North Atlantic centers of knowledge production and circulation. The figures concerning student mobility, for instance, clearly show that the two global leaders of student outward mobility, that is, China and India, prefer the United States and Western Europe as destinations (Varghese 2015, 57; Teplyakov and Teplyakova 2018, 25; Jeanpierre 2010), and that the willingness of students to opt for mobility between China and Brazil is very low (Dwyer et al. 2016).

Measuring the relative importance of BRICS-to-BRICS research collaboration, based on coauthorships as a bibliometric indicator, Finardi and Buratti (2016, 434) assert "very weak" links between BRICS in comparison with their overall international collaboration patterns, and in particular their collaboration with northern partners. Interestingly, the ties between India, Brazil, and South Africa appear relatively stronger. More comparative research on the IBSA (India, Brazil, South Africa) alliance, complementary to or in competition with BRICS, would be needed here.

Beyond institution building and organizing alternative flows of knowledge and people, do the efforts of the BRICS alliance have any epistemic decentering effects? Some participants in the debate around the significance of BRICS for global research and higher education claim that the alliance "is an important attempt to provide an alternative vision of development devoid of the remnants of imperialism and colonialism. The ideas of the Global South, development and interpretations of modernity are then crucial for such collaboration. These ideas lie behind the most developed of the BRICS educational projects, the BRICS Network University" (Khomyakov 2018, 341). This is an ambitious interpretation. It relates, here again, to the mere fact that the NU "will become the largest, the most comprehensive and certainly the most ambitious project as far as South–South cooperation in education is concerned" (Khomyakov 2018, 343). This as such would be an achievement, but it goes back, again, to the creation of decentralized material infrastructures.

The potential to create real alternatives to the existing global higher education landscape has been competing, throughout the development of BRICS higher education and research cooperation, with a more conventional, mainstream take on what is now called excellence. International rankings and benchmarking based on standardized indicators reinforce symbolic centrality. The BRICS have remained ambiguous on how to handle such matters from the beginning and have shifted focus in the course of the years. The initial Russian attempt to introduce comparative benchmarking measures has coexisted with alternative calls for mutual recognition based on South-South solidarity, people-centeredness, and the public good.[11] Incompatible agendas have thus coexisted, and it remains unclear "whether BRICS education cooperation serves the establishment of 'world class universities' for competition in the global higher education market, or whether BRICS focuses on 'national preferences' and on 'common interests and problems of the group rather than global ranking'" (Muhr and Azevedo 2018, 527, quoting David and Motala 2017).

Being Led by States

Decentralization as a political program is carried out, in the case of BRICS, directly by national governments. We can read between the lines and understand from the interviews so far that this strongly state-led agenda does not necessarily enhance the flourishing of more properly intellectual energies. Rather, implicated actors seem to criticize the stifling of academic agendas through government involvement that is too close. The difficulty for the BRICS seems to be to transform a political project into a successful academic one.

This critique, however, requires qualification. Indeed, experiences elsewhere demonstrate that the policy impact on scientific collaboration remains weaker than expected. While none of the official documents refer to the European examples of the Bologna process in the domain of higher education and the EU Framework Programme in the domain of scientific research and innovation, the secondary literature does establish that parallel. The European experience, however, shows that different types of institutions that had benefited from EU funding for their creation and maintenance have in the meantime become established and intellectually recognized institutions. Furthermore, despite the fact that EU funding opportunities are often tied to conditionalities, like having to focus on specific topics considered relevant by the political establishment of the EU, these predetermined objectives that were built into the funding mechanisms have often been only partially achieved. Instead, "by funding the development of an SSH infrastructure, political actors created the conditions for the increasing autonomy of disciplinary fields which, in turn, became able to bend political injunctions in a direction compatible with their specific scientific debates. The extent to which the SSH provide tools to legitimize political processes is, thus, a product of a negotiation between the political and scientific fields" (Heilbron, Boncourt, and Timans 2017, 5). It remains to be seen how far this kind of assessment also holds for other scientific disciplines; and how far this could be a lesson for the future perspectives of BRICS scientific development.

But the example of the coordinated establishment of a European Research Area also teaches us another interesting lesson. Contrary to any expectations, the funding poured into intra-European scientific cooperation has not provoked a networking effect superior to the overall process of transnationalization, since cooperation with non-European countries has developed at a similar pace throughout the considered period. "'Europeanisation' is, in other words, not stronger than the more general trend towards transnationalization. This is a rather surprising finding since intra-European funding and collaboration have markedly increased since the 1990s" (Heilbron and Gingras 2018, 35). This should serve as a caution when it comes to assessing the evolution of cooperative practices, such as in the bibliometric measuring of coauthorships, for example. Apparently, the real-world effect of political agendas, however powerful and endowed with material resources, is not as straightforward.

The Specificities of Megascience

It appears from the analyzed material that important parts of the BRICS cooperation focus on big or megascience, that is, research and technological development that requires extremely costly and locally concentrated scientific infrastructure. The social sciences and humanities disciplines, in turn, remain strongly underrepresented. This, however, introduces a few specific biases into the assessment of the decentralization potential of the projects in question. A quick look at the website of the large-scale telescopes, for example, reveals that their transnational cooperation targets in the majority the usual suspects in the EU and United States. Cooperation within BRICS appears only as a post hoc addition to arrangements that follow the established center-periphery scheme: "This suggests that the bulk of BRICS collaboration in Physics and Astronomy takes place via the medium of international mega-science projects. . . . It turns out . . . that 'collaboration' is currently dominated through the mechanism of mega-science projects that have not been generated through the desire of the BRICS countries to collaborate in S&T" (Kahn 2018, 121). This assessment does not preclude the potential that the newly created mechanisms have to decentralize the existing arrangements. Rather, such qualifications confirm again that we are observing decentralization, that is, the integration of new actors and of new relationships between those actors, rather than the creation of wholesome alternatives or an attempt at delinking.

Conclusion

The academic interactions we have analyzed here are government-led and reflect policy-related areas. As argued above, it is significant that platforms and infrastructures are put in place that undergird such cooperation. The less predictable but fascinating observation is that there is an increase in academic cooperation in the BRICS domain that arises outside the government-led initiatives and reflects a confluence of scholarly interests. In the South African case this is to be found in international relations, sociology, anthropology, development studies, agrarian studies, economics, the fine arts, music, and public health. Some of these initiatives predate the formation of BRICS and were a response to globalization debates and the rise of social movements, especially in the India-Brazil and South Africa nexus. The formation of infrastructure and support for this has strengthened the cooperation, as Africa-linked and South-to-South-linked initiatives are encouraged. There are also mobility grants in the funding mechanisms that are facilitating the movement of academics in each other's domain.

Those developments are beyond the scope of this chapter, but they are relevant, since "there was a belief that the traditional hierarchical structures of the global economy that had characterised the entire twentieth century . . . could now be overcome. There was also trust that this bloc of emerging countries could vocalise the demands of a periphery that was about much more than economic and political exclusion. As the world saw international geopolitics become increasingly multi-polar, this became particularly relevant. The BRICS recognised that they were what centers imagined as periphery, thus deprived of the epistemic devices for creating self-representations" (Pinheiro 2017, 60). If this assessment is adequate, it is clear that the social sciences and humanities disciplines, collaborating across BRICS below the governments' radar, have an important role to play here—alternative self-representations are hardly ever produced in the hard sciences.

Notes

This study was realized within the project After BRICS and New Silk Roads: Reconfiguring the Production and Circulation of Social Science Knowledge between Europe and AfroAsia, under the Gutenberg Chair held by Professor Ari Sitas, University of Strasbourg, 2019–23. The authors would like to thank Tom Dwyer for detailed feedback on earlier versions of our text.

1   See, in particular, Beigel and Salatino (2015); for a global overview, see the last world social science report *Knowledge Divides* (UNESCO 2010); for a critical discussion of recent literature, see Keim (2019).

2   "Taking a closer look at global structures of exchange and communication, however, the predominant pattern is not that of collapsing hierarchies and a 'flattening' universe. Power relations between countries and regions are shifting, established centers are challenged by upcoming ones, but there is little evidence that contemporary social relations would consist of communication flows between more or less equally endowed

individuals, organizations or states" (Heilbron, Boncourt, and Sorá 2018, 2). See also Slaček Brlek, Amon Prodnik, and Thussu 2017.

3 Translations of texts in languages other than English are by the authors.

4 Although several are no longer active, we cite the original websites where we consulted documents in the entries in our "References" list. In some cases, the documents are now available at other websites, but not always in the same format we used when writing this chapter.

5 The section of this chapter on the BRICS Network University is based on an analysis of the following official documents: "Minutes of BRICS Education Ministers and UNESCO Meetings" 2013; BRICS Ministers of Education 2015; BRICS Education Ministers 2015; BRICS Ministers of Education 2016; BRICS Network University 2017; BRICS Ministers of Education 2017, 1; BRICS Network University 2018; BRICS Education Ministers 2018; BRICS Network University 2020; BRICS Ministers of Education 2020. The analysis in the section on STI is based on the following official documents: I BRICS Science, Technology and Innovation Ministerial Meeting 2014; II BRICS Science, Technology and Innovation Ministerial Meeting 2015; MoU on STI Cooperation 2015; III BRICS Science, Technology and Innovation Ministerial Meeting 2015; BRICS Brasil 2019; BRICS Ministers of Education 2020; BRICS RUSSIA 2020, 2020; BRICS Russia 2020, 20. The comprehensive document "STI Overview" (BRICS Russia 2020, 8) was used to complement information on the ministerial meetings for which reports were not available.

6 For a critical discussion, see Khomyakov (2017, 24).

7 The simultaneous project of the BRICS University League has pursued the same intentions but does not seem to have developed at the same pace. The two projects seem to have been launched in parallel, if not in rivalry, but at the same time largely overlapping in terms of participating Chinese and Russian institutions (Lei and Sordia 2017, 111–12; Li 2018, 397–98). The league still has no official website and therefore does not allow for a similar systematic collection of official documents. It was therefore discarded from this analysis. Interestingly, Chinese commentators highlight its significance as "a potential weapon" enabling intercultural exchanges, and as such fitting into Chinese public diplomacy at large, including the option of Chinese funding of the league (Lei and Sordia 2017, 111).

8 See also the analysis by Kiselev and Nechaeva (2018, 59).

9 "Despite of the political incentives for the establishment and deepening of the BRICS STI cooperation, Governments should seek ways to make it sustainable. That means they must create opportunities for scientific communities and enterprises strengthening their interactions and achieving joint concrete results, regardless of the direct involvement of the governments officials. It also means they should foster initiatives whose results can be felt during a long period of time" (BRICS Brasil 2019).

10 Including the National Council of Brazil for Science and Technology Development; Brazil's Innovation Agency; the Ministry of Science and

Higher Education of the Russian Federation; the Russian Foundation for Basic Research (RFBR); the Department of Science and Technology of India; the Ministry of Science and Technology of China; the National Natural Science Foundation of China; the National Research Foundation and the Technology Innovation Agency of South Africa.

11 In 2015, for instance "'sharing' and 'exchanging' of 'best practices' was replaced by 'implementation' of 'international best practices,' while the neoliberal (anglo-centric) policy dimensions of 'benchmarking' and 'excellence' became discursively integrated in November 2015. However, while 'best practice sharing' re-entered the discourse in 2016 . . . the ideas of 'benchmarking' and 'excellence' have been abandoned" (Muhr and Azevedo 2018, 527).

References

Altbach, Philip G. 2012. "The Prospects for the BRICS: The New Academic Superpowers?" *Economic and Political Weekly* 47 (43): 127–37.

Beigel, Fernanda, and Javier Maximiliano Salatino. 2015. "Circuitos segmentados de consagración académica: Las revistas de ciencias sociales y humanas en Argentina." *Información, Cultura y Sociedad* 32: 7–32.

Bertacchini, Enrico, Claudia Liuzza, and Lynn Meskell. 2015. "Shifting the Balance of Power in the UNESCO World Heritage Committee: An Empirical Assessment." *International Journal of Cultural Policy* 23 (3): 331–51. https://doi.org/10.1080/10286632.2015.1048243.

Bornmann, Lutz, Caroline Wagner, and Loet Leydesdorff. 2015. "BRICS Countries and Scientific Excellence: A Bibliometric Analysis of Most Frequently Cited Papers." *Journal of the Association for Information Science and Technology* 66 (7): 1507–13. https://doi.org/10.1002/asi.23333.

I BRICS Science, Technology and Innovation Ministerial Meeting. 2014. "BRICS Science, Technology and Innovation Cooperation: A Strategic Partnership for Equitable Growth and Sustainable Development." Accessed November 26, 2021. http://www.brics.utoronto.ca/docs/140210-BRICS-STI.pdf.

II BRICS Science, Technology and Innovation Ministerial Meeting. 2015. "Brasilia Declaration." Accessed November 26, 2021. http://www.brics.utoronto.ca/docs/150318-sti.html.

III BRICS Science, Technology and Innovation Ministerial Meeting. 2015. "Moscow Declaration: BRICS Science, Technology and Innovation Partnership—a Driver of Global Development." Accessed November 26, 2021. http://www.brics.utoronto.ca/docs/151028-sti.pdf.

VII BRICS Science, Technology and Innovation Ministerial Meeting. 2019. "Campinas Declaration." Accessed November 26, 2021. http:// brics2019.itamaraty.gov.br/images/documentos/Campinas_Declaration_Final.pdf.

BRICS Brasil. 2019. "BRICS Science, Technology and Innovation Work Plan, 2019–2022." VII BRICS Science, Technology and Innovation Ministerial Meeting. http://www.brics.utoronto.ca/docs/190920-BRICS_STI_Work _Plan_2019-2022__Final.pdf.

BRICS Education Ministers. 2015. "III Meeting of the BRICS Education Min- isters Moscow Declaration." Accessed November 22, 2021. http://conf .rudn.ru/conf/nu_brics/2015%20Moscow.pdf.

BRICS Education Ministers. 2018. "The 6th BRICS Education Ministers Meeting Cape Town Declaration on Education and Training." Accessed November 22, 2021. http://nu-brics.ru/media/uploads/filestorage /documents/final_declaration_10_july_2018_approved_by_ministers .pdf.

BRICS Ministers of Education. 2015. "II Meeting of BRICS Ministers of Edu- cation Brasilia Declaration." Accessed November 22, 2021. http://conf .rudn.ru/conf/nu_brics/2015%20Brazil.pdf.

BRICS Ministers of Education. 2016. "New Delhi Declaration on Educa- tion: 4th Meeting of BRICS Ministers of Education." Accessed Novem- ber 22, 2021. http://nu-brics.ru/media/uploads/filestorage/New_Delhi _Declaration.pdf.

BRICS Ministers of Education. 2017. "Beijing Declaration on Education: 5th Meeting of BRICS Ministers of Education." Accessed November 22, 2021. http://nu-brics.ru/media/uploads/filestorage/beijing_declaration _1_1.pdf.

BRICS Ministers of Education. 2020. "Declaration of the 7th Meeting of BRICS Ministers of Education." Accessed November 22, 2021. http://conf .rudn.ru/conf/nu_brics/2020%20Russia%20(online).pdf.

BRICS Ministries of Education. 2015. "Memorandum of Understanding on Establishment of the BRICS Network University." Accessed November 22, 2021. http://nu-brics.ru/media/uploads/filestorage/documents/MoU _SU_BRICS.pdf.

BRICS Network University. 2017. "Zhengzhou Consensus. 2017 BRICS Network University Annual Conference." Accessed November 22, 2021. http://nu-brics.ru/media/uploads/filestorage/zhengzhou_consesus.pdf.

BRICS Network University. 2018. "Third Annual BRICS Network University Conference Declaration." Accessed November 22, 2021. http://conf .rudn.ru/conf/nu_brics/2018%20Stellenbosch_declaration%20NU%20 BRICS.pdf.

BRICS Network University. 2020. "International Governing Board of the BRICS Network University Declaration." Accessed November 22, 2021. http://conf.rudn.ru/conf/nu_brics/4%20Declaration%20IGB%20 NU%20BRICS%20Draft.pdf.

BRICS Russia. 2020. "STI Overview: Five-Year Anniversary of Coopera- tion in Science, Technology and Innovation under the Memorandum of Understanding." Accessed November 26, 2021. https://brics-russia2020.ru /images/113/91/1139196.pdf.

BRICS RUSSIA 2020. 2020. "BRICS STI Declaration 2020." VIII BRICS Science, Technology and Innovation Ministerial Meeting. 2020. Accessed November 26, 2021. http://www.brics.utoronto.ca/docs/201113-sti.pdf.

BRICS STI Framework Programme. 2021. "BRICS Call 2020—Results." February 1, 2021. http://brics-sti.org/index.php?p=new/28.

BRICS STI Framework Programme. n.d. "Projects." Accessed March 29, 2022. http://brics-sti.org/index.php?p=projects,

Coning, Cedric de, and Asha Puri. 2017. "Une volonté partagée de façonner un nouvel ordre mondial." *Hermès, La Revue* 79 (3): 90–96. https://doi .org/10.3917/herm.079.0090.

David, S. A., and S. Motala. 2017. "Can BRICS Build Ivory Towers of Excellence? Giving New Meaning to World-Class Universities." *Research in Comparative and International Education* 12 (4): 512–28. https://doi.org /10.1177/1745499917740652.

Dwyer, Tom. 2017. "Huit ans de travail sur les BRICS." *Hermès, La Revue* 79 (3): 99–106. https://doi.org/10.3917/herm.079.0097.

Dwyer, Tom, Eduardo Luiz Zen, Wivian Weller, Jiu Shuguang, and Guo Kaiyua, eds. 2016. *Jovens universitários em um mundo em transformação: Uma pesquisa Sino-Brasileira*. Brasília: Ipea; Beijing: Social Sciences Academic Press.

Finardi, Ugo, and Andrea Buratti. 2016. "Scientific Collaboration Framework of BRICS Countries: An Analysis of International Coauthorship." *Scientometrics* 109 (1): 433–46. https://doi.org/10.1007/s11192-016-1927-0.

Heilbron, Johan, Thibaud Boncourt, and Gustavo Sorá. 2018. "Introduction: The Social and Human Sciences in Global Power Relations." In *The Social and Human Sciences in Global Power Relations*, edited by Johan Heilbron, Gustavo Sorá, and Thibaud Boncourt, 1–26. London: Palgrave Macmillan.

Heilbron, Johan, Thibaud Boncourt, and Rob Timans. 2017. "Understanding the Social Sciences and Humanities in Europe." *Serendipities, Journal for the Sociology and History of the Social Sciences* 2 (1): 1–9. https://doi.org /10.25364/11.2:2017.1.1.

Heilbron, Johan, and Yves Gingras. 2018. "The Globalization of European Research in the Social Sciences and Humanities (1980–2014): A Bibliometric Study." In *The Social and Human Sciences in Global Power Relations*, edited by Johan Heilbron, Gustavo Sorá, and Thibaud Boncourt, 29–58. London: Palgrave Macmillan.

Huet, Sylvestre. 2018. "La place de la France dans la science mondiale." *Le Monde*, April 5. https://www.lemonde.fr/blog/huet/2018/04/05/la-place -de-la-france-dans-la-science-mondiale/.

Jeanpierre, Laurent. 2010. "The International Migration of Social Scientists." In *World Social Science Report 2010*, edited by International Social Science Council, 118–21. Paris: UNESCO.

Kahn, Michael. 2015. "Prospects for Cooperation in Science, Technology and Innovation Among the BRICS Members." *International Organisations*

Research Journal 10 (2): 105–19. https://doi.org/10.17323/1996-7845-2015 -02-140.

Kahn, Michael. 2018. "Co-authorship as a Proxy for Collaboration: A Cautionary Tale." *Science and Public Policy* 45 (1): 117–23. https://doi.org/10 .1093/scipol/scx052.

Keim, Wiebke. 2011. "Counter Hegemonic Currents and Internationalization of Sociology: Theoretical Reflections and One Empirical Example." *International Sociology* 26 (1): 123–45.

Keim, Wiebke. 2019. "Les savoirs des sciences sociales à l'international: Séquences empiriques, débats théoriques, enjeux épistémologiques." Habilitation à Diriger des Recherches, Sociologie, Université Paris V—René Descartes. https://hal.archives-ouvertes.fr/tel-02481741/document.

Khomyakov, Maxim. 2017. "Towards Sustainable BRICS Collaboration in Education." *China Today*, September 15. http://www.china.org.cn /opinion/2017-09/15/content_41593777.htm.

Khomyakov, Maxim B. 2018. "BRICS and Global South: Towards Multilateral Educational Collaboration." *Changing Societies and Personalities* 2 (4): 329–50. https://doi.org/10.15826/csp.2018.2.4.050.

Kiselev, Vladimir, and Elena Nechaeva. 2018. "Priorities and Possible Risks of the BRICS Countries' Cooperation in Science, Technology and Innovation." *BRICS Law Journal* 5 (4): 33–60. https://doi.org/10.21684/2412-2343 -2018-5-4-33-60.

Lei, Wang, and Caroline Sordia. 2017. "Échanges interculturels et ligue des universités des BRICS: Le point de vue de la diplomatie publique chinoise." *Hermès, La Revue* 79 (3): 111–13. https://doi.org/10.3917/herm .079.0111.

Li, Yuyun. 2018. "Development of Cooperation in Higher Education in BRICS Countries." *Changing Societies and Personalities* 2 (4): 393–405. https:// doi.org/10.15826/csp.2018.2.4.053.

Mignolo, Walter D. 2012. "The Role of BRICS Countries in the Becoming World Order: 'Humanity,' Colonial/Imperial Differences, and the Racial Distribution of Capital and Knowledge." In *Humanity and Difference in the Global Age*, 41–89. UNESCO and Universidad Candido Mendes, Brazil.

"Minutes of BRICS Education Ministers and UNESCO Meetings." 2013. Accessed November 22, 2021. http://conf.rudn.ru/conf/nu_brics/2013%20 BRICS-UNESCO.pdf.

MoU on STI Cooperation. 2015. "Memorandum of Understanding on Cooperation in Science, Technology and Innovation between the Government of the Federative Republic of Brazil, the Russian Federation, the Republic of India, the People's Republic of China and the Republic of South Africa." Accessed November 26, 2021. http://www.brics.utoronto.ca/docs /BRICS%20STI%20MoU%20ENGLISH.pdf.

Muhr, Thomas, and Mário de Azevedo. 2018. "The BRICS Development and Education Cooperation Agenda." *Vestnik RUDN. International Relations* 18 (3): 517–34. https://doi.org/10.22363/2313-0660-2018-18-3-517-534.

Oustinoff, Michaël. 2017. "Introduction: Les BRICS, un espace ignoré." *Hermès, La Revue* 79 (3): 13–18.

Pinheiro, Cláudio Costa. 2017. "The BRICS Countries: Time and Space in Moral Narratives of Development." In *The Moral Mappings of South and North*, edited by Peter Wagner, 51–71. Annual of European and Global Studies. Edinburgh: Edinburgh University Press.

Shashnov, Sergey, and Maxim Kotsemir. 2018. "Research Landscape of the BRICS Countries: Current Trends in Research Output, Thematic Structures of Publications, and the Relative Influence of Partners." *Scientometrics* 117 (2): 1115–55. https://doi.org/10.1007/s11192-018-2883-7.

Sitas, Ari. 2006. "The African Renaissance Challenge and Sociological Reclamations in the South." *Current Sociology* 54: 357–80.

Slaček Brlek, Sašo, Jernej Amon Prodnik, and Daya K. Thussu. 2017. "'Is Enlightenment Just a European Idea?' An Interview with Daya Thussu." *tripleC* 15 (1): 285–304. https://doi.org/10.31269/triplec.v15i1.862.

Sorokotyaga, Yaroslav. 2021. "BRICS STI FP—5 Years!" BRICS STI Framework Programme, May 3. http://brics-sti.org/index.php?p=new/29.

Teplyakov, Dmitry, and Olga Teplyakova. 2018. "National Policy for Academic Mobility in Russia and the BRICS Countries: 20 Years of the Bologna Process Implementation." BRICS *Law Journal* 5 (1): 5–26. https://doi.org/10.21684/2412-2343-2017-5-1-5-26.

Thussu, Daya, and Françoise Wirth. 2017. "Internet des BRICS et désoccidentalisation des sciences de la communication." *Hermès, La Revue* 79 (3): 65–70. https://doi.org/10.3917/herm.079.0065.

UNESCO. 2010. *World Social Science Report: Knowledge Divides*. Paris: UNESCO; International Social Science Council.

UNESCO. 2014. "BRICS—Building Education for the Future. Priorities for National Development and International Cooperation." Paris: United Nations Educational, Scientific and Cultural Organization.

Varghese, N. V. 2015. "BRICS and International Collaborations in Higher Education in India." *Frontiers of Education in China* 10 (1): 46–65. https://doi.org/10.1007/BF03397052.

Zyw Melo, Anna. 2017. "Un câble pour les BRICS: Un défi stratégique insurmontable." *Hermès, La Revue* 79 (3): 145–49. https://doi.org/10.3917/herm.079.0145.

Contributors

LINDA MARTÍN ALCOFF is Professor of Philosophy at Hunter College and the Graduate Center, CUNY. She is a past President of the American Philosophical Association. Her areas of work include epistemology, Latin American philosophy, feminism, critical race theory, and continental philosophy. Her books include *Rape and Resistance* (2018), *The Future of Whiteness* (2015), and *Visible Identities: Race, Gender and the Self* (2006), which won the Frantz Fanon Award. She has also edited or co-edited eleven books and written over one hundred journal articles and book chapters, and has contributed to the *New York Times, Aeon, NY Indypendent,* and other publications. She is working on two books: one on a decolonial approach to race and racism, and a second on extractivist epistemologies. She is originally from Panama.

ELÍAS BARTICEVIC is a journalist and social scientist from the Chilean Magallanes Region with much interest and experience in science communication and social valuation. He has participated in research around the relationships between science, society, and cultural development as related to Antarctica. He is currently head of the Competitive Projects section of the Chilean Antarctic Institute.

JOHAN HENRIK M. BULJO is a well-known Sámi knowledge holder. He has fished and hunted, picked berries, and fetched winter wood since he was ten. He is now seventy-six. He calls *meahcci* his home, and he has a close relationship with and a great respect for it. For him, *meahcci* is a greater actor than any human being. It has its own will. Buljo is also very concerned about sustainability and that the land and the waters, animals, birds, and plants should not be vandalized with mines, windmills, pollution, or other new technologies. He has deep knowledge about the ducks and geese in Sápmi. Although he hunts ducks in the Sámi season of winter-to-spring, he respects and has consideration for the birds. He was the leader of the Lodden Committee in Guovdageaidnu that drafted the report described in his contribution to this volume.

RONALD CANCINO holds a PhD in social sciences from the Universidad de Chile. He is an academic in the Department of Social Sciences at the Universidad de la Frontera and Coordinator of the Research Group on Science, Technology, Society, and Territory. His academic interests focus on problems of what he has called "socio-scientific complexity," an interweaving of relationships between scientific, technological, social, and cultural phenomena that are activated around sociotechnical controversies, public policy dilemmas, disputes between networks, and types of expert and local knowledge. He is also interested in modeling methodologies of science as a complex adaptive system to analyze explanatory mechanisms of science as an emerging phenomenon, as well as the understanding of scientific capabilities. He is President of ESOCITE, Latin American Society for Social Studies of Science and Technology, and researcher in the CYTED Project Science, Technology and Innovation Policies, Transferable Models at a Local Scale.

CRISTINA FLORES is an Adjunct Professor at Universidad Nacional de la Patagonia Austral (UNPA). She holds a BA in education from the National University of Quilmes (UNQ). She is currently a PhD candidate in economic development at UNQ.

KIM FORTUN is Professor of Anthropology at University of California, Irvine, and studies environmental risk and disaster. She is an interdisciplinary, mixed-methods ethnographer specializing in comparative studies of environmental knowledge, injustice, and governance. She teaches environmental studies, science and technology studies, and experimental ethnographic methods and research design. She uses experimental ethnographic methods to understand how people in different geographic regions and organizations deal with environmental problems, health risks, and major disasters. Fortun is particularly focused on industrial disasters—chemical plant explosions and massive breakdown of industrial systems. Her fieldwork has spanned highly populated regions in India and the United States, where she has focused on both catastrophic disasters and slower-moving disasters, including air pollution.

SANDRA HARDING is a Distinguished Research Professor Emerita at UCLA. She has authored or edited seventeen books and special journal issues on topics in feminist and postcolonial epistemology, philosophy of science, and methodology. She coedited *Signs: Journal of Women in Culture and Society* from 2000 to 2005. She has consulted with the Pan American Health Organization, UNIFEM, the UN Commission on Science and Technology for Development, and UNESCO. She has held visiting appointments at the University of Costa Rica, the University of Amsterdam, the Swiss Federal Institute of Technology Zurich (ETH), and the Philosophy Department of Michigan State University. She is cofounder and Senior Advisor Emerita of the journal *Tapuya: Latin American Science, Technology and Society*.

LINE KALAK is a Sámi lawyer, academic, and politician. She engages with and is an advocate of Sámi rights, especially the right of Sámi people to live their culture and to use and to manage their own lands and resources. Sámi traditional knowledge related to land and resources is one of Kalak's core concerns, and in 2020 she published a book about Sea Sámi salmon fishers and their traditional knowledge. In 2019 she published an article on the legal protection of coastal Sámi culture and livelihood in Norway. She is also a former Vice Rector of the Sámi University of Applied Sciences, where she currently works as an Assistant Professor of Law.

DUYGU KAŞDOĞAN is Assistant Professor of Urbanization and Environmental Problems in the Department of Political Science and Public Administration at İzmir Katip Çelebi University, Turkey. She received her doctoral degree in the Science and Technology Studies Program at York University, Canada. She was a visiting researcher in the Anthropology Department at MIT in 2013–14, and a research fellow in the Sociology Department at Koç University under the TÜBİTAK Co-funded Brain Circulation Scheme fellowship program in 2016–17. She is the founding member of IstanbuLab and Transnational STS Network, and associate editor of the journal *Engaging Science, Technology, and Society*. Her research focuses on democratization of science, transnational collaboration, political ecology of disasters, toxicity governance, and bioeconomies.

WIEBKE KEIM has been employed as a CNRS researcher at SAGE (Sociétés, Acteurs, Gouvernement en Europe), Strasbourg University, since 2013.

Publications include "Vermessene Disziplin: Zum konterhegemonialen Potential afrikanischer und lateinamerikanischer Soziologien" (2008); "Global Knowledge Production in the Social Sciences: Made in Circulation" (with Ercüment Çelik, Christian Ersche, and Veronika Wöhrer, 2014); "Gauging and Engaging Deviance, 1600–2000" (with Ari Sitas, Sumangala Damodaran, Nicos Trimikliniotis, and Faisal Garba, 2014); "Universally Comprehensible, Arrogantly Local: South African Labour Studies from the Apartheid Era into the New Millennium" (2017); "Scripting Defiance: Four Sociological Vignettes" (with Ari Sitas, Sumangala Damodaran, Amrita Pande, and Nicos Trimikliniotis, 2022); and *Handbook on Academic Knowledge Circulation* (editor-in-chief with Leandro Rodriguez Medina, 2023). Her focus areas are history, sociology, and epistemology of the social sciences, circulation of knowledge, and fascisms.

AALOK KHANDEKAR is a science and technology studies scholar and an Assistant Professor of Anthropology/Sociology in the Department of Liberal Arts and Affiliated Faculty at the Department of Climate Change at the Indian Institute of Technology Hyderabad (IITH). In recent years, two distinct but related concerns have been at the center of his scholarly attention: first, a concern with environmental and climate change governance in southern city settings; second, understanding and creating the conditions for interdisciplinary collaborations addressing urgent challenges, such as those presented by global climate change. Increasingly, much of his research has assumed a transnational character, where he has worked to establish and sustain networks of scholarship that are in conversation about and offer comparative perspective on shared concerns. Khandekar also serves as the editor in chief of *Engaging Science, Technology, and Society*, the diamond open-access journal of the Society for Social Studies of Science (4S).

DANIEL LEE KLEINMAN is Professor of Sociology, Boston University. He served as Associate Professor for Graduate Affairs at BU from 2017 through 2024. Before coming to BU in January 2017, Kleinman spent seventeen years on the faculty at the University of Wisconsin–Madison. Broadly speaking, Kleinman's scholarly research explores the politics of science. Some of this work focuses on politics in the colloquial sense—that is, he is interested in science policy. In other research, the word *politics* highlights Kleinman's interest in the social organization of power and its institutionalization as factors that affect how knowledge is developed. The arc of Kleinman's scholarship across nearly twenty-five years fits into two broad categories:

the sociology of the knowledge economy and academia and the sociology of democracy and expertise.

WEN-HUA KUO is a Professor at National Yang Ming Chiao Tung University, Taiwan, where he teaches social studies of medicine. A licensed physician, his work revolves around pharmaceutical regulation and its social impacts in the East Asian context, and controversies in attempts to modernize and use East Asian medicines globally. His scholarly publications appear in a range of journals crossing several disciplines, including the *Journal of Law, Medicine, and Ethics*; *Drug Information Journal*; *Social Science and Medicine*; *Science, Technology and Society*; and *Isis*. He has also contributed chapters in the books *Lively Capital* and *Global Health and the New World Order*. In addition to his current research on care and caring professions, he served as the editor-in-chief of *East Asian Science, Technology and Society: An International Journal* from 2016 to 2022, and has served as one of the associate editors of *Social Studies of Science* since 2023. He was elected and will serve as President of the Society for Social Studies of Science (4S) from 2025 to 2027.

JOHN LAW is Emeritus Professor of Sociology at the Open University in the UK. He has worked in science and technology studies on actor network theory, material semiotics, and their successor projects, and most recently on postcoloniality and on less colonial ways of knowing. Here he has coauthored with Sámi scholars and activists on indigenous struggles in Norway over fishing, hunting, land use, and language, and with Taiwanese STS scholar Wen-yuan Lin on a less colonial social science inflected by explanatory logics drawn from Chinese medicine. He has also written with Dutch empirical philosopher Annemarie Mol on the consequences of the domination of the English language for social science. He is the author of numerous journal articles, book chapters, and (edited) books including *After Method: Mess in Social Science Research* (2004; translated as *Después del método: Desorden en la investigación en ciencias sociales*, 2020) and a volume coedited with Annemarie Mol, *On Other Terms: Social Theory and Non-English Languages*, published in 2020 as Sociological Review Monograph 68 (2). His personal web page is Heterogeneities.net (http://www .heterogeneities.net).

LES LEVIDOW is a Senior Research Fellow at the Open University, UK, where he has studied agri-environmental-technology issues, especially techno-market fixes, consequent controversy, and alternative agendas. A longtime case study was the agri-biotech (transgenics) controversy, focusing on the European Union, the United States, and their trade conflicts. This research resulted in two books: *Governing the Transatlantic Conflict over Agricultural Biotechnology: Contending Coalitions, Trade Liberalisation and Standard Setting* (2006), and GM *Food on Trial: Testing European Democracy* (2010). His most recent book is *Beyond Climate Fixes: From Public Controversy to System Change* (2023). Over the past decade, his research themes have broadened to agricultural innovation priorities, the bioeconomy, corporate environmental stewardship, natural capital assessment, alternative agrifood networks, community gardens, and the agroecology-based solidarity economy in South America. Details are available at his web page (https://fass.open.ac.uk/people/ll5). As an early entry point into such issues, he participated in the Radical Science Journal Collective in the 1970s–80s. It was replaced by a new journal, *Science as Culture*, for which he has served as coeditor.

ANGELA OKUNE studies and works on scholarly research infrastructures including open data, equity in open science, and open access publishing. Trained as a social scientist of science, she received her doctorate in Anthropology from the University of California, Irvine, and has been awarded research fellowships by the National Science Foundation, the Wenner-Gren Foundation, and the University of California, Berkeley, Center for Technology, Society and Policy. From 2010 to 2015, as cofounder of the research department at iHub, Nairobi's innovation hub for the tech community, Angela provided strategic guidance for the growth of tech research in Kenya. She worked as a Community Coordinator for the Open and Collaborative Science in Development Network (OCSDNet; 2014–18) and coeditor for the open-access book *Contextualizing Openness: Situating Open Science* (2019). Angela serves as a Design Team member of the Platform for Experimental Collaborative Ethnography (PECE) and as an Associate Editor for the open-access journal *Engaging Science, Technology, and Society*. She also founded an experimental, open ethnographic data portal called Research Data Share.

LIV ØSTMO was one of the founders of the Sámi Allaskuvla/Sámi University of Applied Sciences, Guovdageaidnu, Norway, and is a former Dean. Now retired, she has lectured and undertaken research on multicultural understandings. Over the last decade, she has worked to document traditional

Sámi environmentally relevant knowledge in a range of contexts, in articles, reports, and films. She is also an activist, and has been struggling for Sámi rights for decades. She took a leading role in the protest against the Guovdageainnnu-Alta River hydroelectric project in the 1970s and early 1980s.

LEANDRO RODRIGUEZ MEDINA is Professor of Science, Technology and Society and Director of the Center for the Study of Science, Technology and Society at Universidad Alberto Hurtado (Chile) and a National Researcher (Level II) at the Mexican Council for the Humanities, Science, and Technology (CONAHCYT). He holds a PhD in sociology from the University of Cambridge, and his research interests are the international circulation of academic knowledge, the relationship between culture and urban development, and the social aspects of infectious diseases. Founding editor-in-chief of *Tapuya: Latin American Science, Technology and Society* (2023 Infrastructure Award from the Society for Social Studies of Science), his books include *Centers and Peripheries in Knowledge Production* (2014), *The Circulation of European Knowledge: Niklas Luhmann in the Hispanic Americas* (2014), and, as editor, the *Routledge Handbook of Academic Knowledge Circulation* (2023) and *La Teoría del Actor Red en América Latina* (2022). He has served as Associate Editor of *Social Studies of Science* since 2023 and held visiting positions in France, Germany, and Mexico. He chaired the 2014 (Buenos Aires) and 2022 (Cholula, Mexico) joint meetings of the Society for the Social Studies of Science and the Latin American Association for Social Studies of Science and Technology (ESOCITE). Forthcoming books include an introduction to STS, a monograph about Mexico's midsized cities as cultural urban assemblages, and an edited volume on Latin American theoretical contributions to STS.

ARI SITAS is an Emeritus Professor of Sociology at the University of Cape Town and the Director of the Ministerial Team for the Charter of the Humanities and Social Sciences; he is also a Fellow at the Institute of Advanced Studies, Jawaharlal Nehru University, New Delhi, and a Guest Professor at the Albert-Ludwigs University of Freiburg. He joined UCT as a professor in May 2009, after twenty-six years at the University of Natal and later the University of KwaZulu-Natal. He has been a senior fellow and research associate in a number of institutions: the University of California, Berkeley; Ruskin College; and Oxford University. He is past President of the South African Sociological Association, Vice-President of the

International Sociological Association, and executive member of the African Sociological Association. He completed his PhD at the University of the Witwatersrand on the emergence of trade unions and social movements among Black urban and migrant workers (1960s–1980s) under the supervision of Eddie Webster and the late David Webster in 1984. Sitas is also a writer, dramatist, and poet and has been a leading intellectual and an activist in the anti-apartheid movement.

MAKA SUAREZ is an Assistant Professor in the Department of Social Anthropology at the University of Oslo. Her work moves across political, economic, and multimodal anthropology. Her book manuscript is about the lives of Latin American migrants who reoriented political mobilizations in Europe in the face of excessive debt and home eviction. Following housing struggles and creative forms of collective organizing against indebtedness in Spain for the past decade, she ethnographically documents the making of transnational financial capitalism and the potential for emancipatory democratic transformation. She is also developing new research projects at the interfaces of socioecological knowledge cocreation, science and technology studies, and the potential of more-than-textual spaces for decolonial hermeneutics—such as EthnoData, a platform she coproduced to build accessibility to data with the Kaleidos Center for Interdisciplinary Ethnography at the University of Cuenca, Ecuador, which she cofounded.

SHARON TRAWEEK is an Associate Professor in the Department of Gender Studies and History at UCLA; she has also been on the faculty of the Anthropology Department at Rice University and the Program in Anthropology and Archaeology and the Program in Science, Technology, and Society at MIT. She has held visiting faculty positions at universities in Japan, Sweden, and the United States. She received her PhD from the History of Consciousness Program at the University of California, Santa Cruz. Her first book is *Beamtimes and Lifetimes: The World of High Energy Physicists* (1988); it remains in print and has been translated into Chinese (2003). Her most recent book, coauthored with Knut Sørensen, is *Questing Excellence in Academia: A Tale of Two Universities* (2022), and her next is on epistemic edges. She has also published articles and chapters in books and journals of anthropology, Asian studies, communications, cultural studies, history, and gender studies. In 2020 she was awarded the J. D. Bernal Prize for "distinguished contributions to the field of STS" from the interdisciplinary, international Society for Social Studies of Science (4S).

HEBE VESSURI is an Argentine-Venezuelan social anthropologist who has worked in different countries, spending more or less long periods in Argentina, Brazil, Canada, Colombia, France, Germany, England, Mexico, and Venezuela. An important part of her career was linked to international scientific cooperation, holding positions, among others, at the United Nations University in Tokyo, UNESCO, IHDP, and ICSU. Her research interests have varied over time, although recently they are focused on the production and distribution of knowledge in globalization, how social knowledge accompanies the transformations of contemporary societies, expert knowledge, and the crisis of social critique. In 2018 she received the Argentine government's Bernardo Houssay Award for Scientific Career Achievement in the Humanities; in 2017 the John Desmond Bernal Award for Distinguished Contribution in the field of Social Studies of Science and Technology, awarded in Boston by the Society for the Social Studies of Science (4s); and in 2014 the Oscar Varsavsky Career Achievement Award from the Latin American Society for the Social Studies of Science and Technology (ESOCITE). She is a researcher emeritus at IVIC (Venezuelan Institute for Scientific Research).

Index

cancer control and care, 225, 233

canonization in Machu Picchu, 76–79

Caporal, Francisco Roberto, 191

Caring to Know (Dalmiya), 37

Castro, Eduardo Viveiros de, 48, 49

CENPAT (Centro Nacional Patagónico) (Argentina), 252

centers of knowledge production, 22–23n2

centralizing institutions, 147

ceremonial objects, 47

Chacareros General Carrera Lake (Argentina; Chile), 243

Chakrabarty, Dipesh, 113

Chao, Co-Shi Chantal, 222–23, 224–25, 227, 231–32, 235n4

chemical colonialism, 205

Chile: climate change turn, 253; fire monitoring system, 246–47; mining, extractivism, and national policies, 243, 256n1; salmon farming and contamination, 244–45; scientific capacities, 248–49; socioterritorial turn in scientific concerns, 251, 252; STI policy, 253, 254

Chilean astronomy and epistemic decentralizing, 79–83

Chilean pine (*Araucaria araucana*), 247

China, 261, 262; Google in, 276. *See also* BRICS (Brazil, Russia, India, China, and South Africa)

Chinese medicine, 227, 231, 236n16

chiropractic methods, 236n12

Chung, Chang-Hung, Dr., 235n6

circuitos curtos (short supply chains), 187–88, 198, 199, 200, 201, 203, 204

#CiteBlackWomen, 10

citizen science, 4, 75

Civil Society for Climate Action (Chile), 243

classifications, building trust outside the, 119–20

Clifford, James, 45–46

climate change project, 158

climate change turn, 253

coalition building, 43, 46

Code, Lorraine, 41

collaboration: knowledge production through, 15, 48–49, 53–54, 102, 158; research, 271–73

collaborative nature of human knowledge. *See* neem patenting case

collaborative versus extractive approaches, 39–44

collective marketing: of agroecological products, 203; cooperative skills for, 197; promoting cooperatives for, 204

Colombia, 33

colonial academic logics, 154

colonialism: chemical, 205; and epistemic extractivism, 6, 15; through patent rights, 39, 40–42. *See also* metropoles; Sámi resistance

colony collapse disorder (CCD), 99, 101

comfort care, 234n3

comfort nursing, 223, 231–32, 235n4, 236n13

comida de verdade (real food), 199, 203

Comisiones Convención Constitucional (Chile), 256n1

commodification, 114

communication practices, 48–49, 110, 111, 118, 119, 120

community beliefs, 117

complementary and alternative medicines (CAM), 227, 236n12

Comunidades Eclesiais de Base (CEBS) (Brazil), 193

conformism, 53

Connell, Raewyn, 74

conscientious consumers, 199

Consejo Nacional de Investigaciones Científicas y Técnicas (CONICET) (Argentina), 249, 252

conservation of land: fire and megafire risks to, 246–48

contested conviviality, 167

contextualization/decontextualization/recontextualization, 45, 49, 67

contract funding, 158

contract time, 162–64, 170

Control Yuan, 232–33

cooperative skills for collective marketing, 197

cooperativism, 204

Cooperativismo e Soberania Alimentar (Brazil), 204

corridor talk, 117, 124

corroboration, 117, 118

Costabeber, José Antônio, 191

costs, 160, 161, 176n15

Council for Structural Change (Argentina), 246

electronic acupuncture, 227, 228

elites, 153–54, 161

Elliot, Kevin C., 95

Elsevier (publisher; aka Reed-Elsevier), 160, 162, 175n13

embodied knowing, 134–38, *145*

emigration, 23n3

empires, 2, 5–6, 50, 68, 113, 116, 175n6. *See also* colonialism; metropoles; periphery

empowerment, 8, 62, 71, 83–84, 189, 240

Empresa Brasileira de Pesquisa Agropecuária (Embrapa), 194

end-of-life care, 224, 235n10. *See also* quality of death and dying, study on

energy treatments, 229, 236n12

Engaging Science and Technology Studies (journal), 166

environment: disaster, 172; justice, 102–5; and knowledge, 102–5, 134–38; and oppression, 133–34

Environmental Protection Agency (United States), 101

epistemic systems: agency and, 7, 185; authority and, 74, 112, 115; centralized, 3, 5–6, 15, 17, 18, 69, 185; and coloniality, 23n3; and co-opted knowledge, 69–72; deauthorization of knowledge, 44; decentering of ideas, 3, 5, 7; decentralization, 4–5, 275, 278; Atacama Desert (Chile), 79–83; definition, 4, 5–8; imbalance in, 23n3; innovations in, 275; Latin American agroecology, 191–93; Machu Picchu (Peru), 76–79; pluralism in, 41

epistemic extractivism. *See* extractivist epistemology

epistemic forms, 95, 96, 99–100, 101, 104–5

epistemic humility, 12, 15, 16, 37, 47, 50, 62, 76

epistemicide, 185

epistemic indifference, 63–66

epistemic injustice, 36, 40, 67, 74. *See also* neem patenting case

epistemic methods, 34–35

epistemic mutualism, 16, 62, 63, 72–76, 84–85n3

epistemic-organizational shift, 193–95

epistemic power relations, 256, 264

epistemic practices, 34–35; corrective norms for alternative, 50–57; at the edge, 116–19; multiple, overlapping, 114–16

epistemic privilege, 109; authority and, 120–21; challenging established practices, 121–22; classifications and, 119–20; practices at the edge of, 116–19; exclusionary practices, 121; imperial and colonial, in global academia, 113–14; manifestations in academic discourse, 112–13; in multiple, overlapping practices, 114–16; traditional academic writing and narrative strategies, 122–25

epistemic reliability, 46

epistemic status of marginal voices, 111

epistemological assumptions, 137, 147

epistemology: concept of, 57n1; of ignorance in patent claims, 39–44

Epstein, Steven, 97–99

equivocations, 48–49, 78

Escobar, Arturo, 39–40

essentialism, 113, 116, 122

essential oils, 228

ethical issues in hospices, 236n19

ethnobotany, 44

ethnosciences, 115

eucalyptus, 228

EU Framework Programme, 279

European museums, 47

European Research Area, establishment of, 279

exceptionalisms, 54, 113

exclusionary practices, 121

experiential knowledge, 124

extractivism: as an epistemology, 34–38; and national policies, 243–44; practice and idea of, 32–34

extractivist capitalism, 32

extractivist epistemology, 34–38, 66–69, 184–85; case studies, 38–49; corrective norms for alternative epistemic practices, 50–57; epistemic norms counteracting, 50–51

false negatives, 101

false positives, 101

Family Agriculture Secretariat (Brazil), 204

faultlines, 109, 122

feiras do agricultor (farmers' markets) (Brazil), 199, 204

feiras livres (outdoor markets) (Brazil), 198

feminist epistemologies, 112–13, 114–15, 124

feminist epistemologists, 123
feminist standpoint theory, *10*
Feyerabend, Paul, 51
Finkelstein, Eric, 222
Finnmark (Norway), 133
fire and megafire risks to biodiversity, 246–48
FLACSO-Ecuador (Latin American Faculty of Social Sciences–Ecuador), 167
Fome Zero (Zero Hunger) campaign (Brazil), 196
food safety and quality, 202
Fortun, Kim, 125, 156
Fortun, Mike, 125
Forum Brasileiro de Economia Solidária (FBES), 197
Foucault, Michel, 131–32
fractionalities, 73–74
Fraser, Nancy, 9–10
Freidson, Eliot, 234
Frente Ecológico Austral Defensa (Chile), 243
Frickel, Scott, 95
Fricker, Miranda, 9, 12
Fundación Terram (Chile), 243, 245
funding: agencies' operational overhead, 176n15; BRICS Network University, 268; collective funding mechanism, 271; by the EU, 279; iHub (Kenya), 163; internationalization in Indian academic institutions, 156, 157, 158; intra-European scientific cooperation, 279; multilateral joint projects, 270; PRONAF (Brazil), 190; public universities in Ecuador, 160; for scientific research, 248; shortage, 169. *See also* New Development Bank of (BRICS)

Gabriel García Márquez Archive, 68
Gandhi, Mohandas Karamchand, 34
General Law of Fisheries and Aquaculture of Chile, 244
geographically decentralized academia: Ecuador, academic publishing in, 160–62; India, internationalization in, 156–59; Kenya, coloniality in, 162–64. *See also* academic decentralization
GIAN (Global Initiative for Academic Networks) (India), 175n8
Gikuyu (language), 174n3
global design, 6

Global North, 31, 33, 40, 80, 161, 185, 194
Global South, 7, 13, 22n1, 42, 158, 190; academics and publishers, 161; development and interpretations of modernity, 278; marginalized research, 261; universities, 153–54
gnoseology, 52
"gold standard" in research, 117
Google, 276
gossip, 110, 111, 118, 119, 120
government-to-government initiatives, 276
Greene, Shane, 44
Greenpeace, 243, 245
Green Revolution, 194
grief counseling for hospice patients, 236n19
Gross, Matthias, 94
group relations, 45–46
Guevara, Perry, 173n2
Guiren, Yuan, 265
Guovdageaidnu Council (Norway), 138, 139, 148–49

Hamuy, Mario, 79
Han, Ji Sheng, 229
Hanaoka, Seishu, 236n14
harassment, 120
Haraway, Donna, 94, 102–3
Harding, Sandra, 54, 55, 93, 94, 102–3, 115, 155, 172
Harrison, Jill Lindsey, 105
Hayden, Cori, 42
healing traditions, 227, 229–30, 236n11
Healthcare Model Award, 235n6
hegemony-seeking epistemologies, 54
Henry, Jade Vu, 125
herbal remedies, 195, 227, 228, 236n12, 236n16
hermeneutic injustice, 9, *10*, 12
hermeneutics, 52
Hernandez, Ariel, 125
Herzig, Rebecca, 176
Hess, David, 95
hierarchies: of epistemic authority, 115; in imperial metropoles, 114; of state-friendly infrastructures, 140, 147
higher education policy, 263, 264, 265–67, 275, 276, 277, 278, 279
h-indexes, 157, 175n9
Holbrook, Jarita, 125
homo economicus model, 187

homo situs, concept of, 201
honeybee health, 93, 99–101
hospice, concept of, 235n6
Hospice Foundation of Taiwan, 225, 232
hospice practices (Taiwan): decentralizing
    biomedicine in, 226–30; Hospice Home
    Care Benefits and Subsidies, 225; Hospice
    Palliative Care Regulations, 222; Hospice
    Palliative Medical Regulations, 235n7;
    policies, 222–23; shared-care scheme,
    235n5
hospice wards (Taiwan): operation of, 226;
    in national health care system, 224–26; as
    therapeutic and knowledge space, 221–24,
    226–30
human remains, as artifacts, 45–49
humor, 117
hygiene restrictions, 202

ignorance, 94, 96–102, 104; AIDS treatment
    activism, 97–99, 103; beekeeper knowl-
    edge making, 99–102, 103; Bucket Brigade
    case, 102–5; meaning of, 94–96; and the
    value of decentered and decentralized
    knowledge production, 96–102
iHub (Kenya), 162–64
imperial metropoles, 113–14
import substitution industrialization, 33
*Improvised News: A Sociological Study of
    Rumor* (Shibutani), 119
Inca people, 62, 77, 78. *See also* Machu
    Picchu (Peru)
incompleteness of knowledge, 51–52
Incubadora de Tecnologia Social (ITS)
    (Brazil), 198
India, 175n8; academic institutions, 155,
    156–59; people of, 47. *See also* BRICS
    (Brazil, Russia, India, China, and South
    Africa)
Indian Agriculture Research Institute, 42, 55
indigenous epistemologies, 39–44
indigenous knowledge, 134–38, *145*
indigenous knowledge and biopiracy: corpo-
    rate acquisition, 39–40; bio-prospecting,
    42; case study: neem tree, 42–44; patent
    rights, 39, 40–42; core concepts, 40;
    "Scientific/Intellectual Movements"
    (SIMS), 40
Indigenous Knowledge Systems (IKS), 115

indigenous peoples: intellectual property
    rights, 125n2; in Zinacantán, 71. *See also*
    specific tribes
indigenous practices, 138. *See also lodden*
    (Sámi spring-winter duck hunt)
indigenous resistance, 133–34, 146–50
individualism, 53, 55
inductive and deductive reasoning, 112
infectious salmon virus (ISA), 244–45
informal knowledge, 118
Information Centre at the University of
    Toronto, 264
infrastructural resources, 137
infrastructure of knowledge, 5, 14, 18, 35,
    68–70, 79, 143, 155–56, 162, 163, 164, 166–67,
    169, 170, 171, 176, 254. *See also* Atacama
    Desert (Chile); research infrastructure
infrastructures of knowing, *145–46*; embod-
    ied knowing, 134–38, *145*; textual knowing,
    139–43, *145*, 147–48
insider knowledge, 37, 56
Instituto de Tecnologia Social (Brazil), 198
intellectual property rights, 125n2. See also
    World Intellectual Property Organization
intercultural translation, 78
inter-epistemology, 51–53
internationalization in Indian academic
    institutions, 155, 156–59
international thematic groups (BRICS Net-
    work University), 268
intersectionality, *10*
invisible knowledge spaces in Taiwan's pallia-
    tive care, 221–24
iron cage, 187
Isla Grande of Chiloé (Chile), crisis on, 245
IstanbuLab, 18, 165–67

joint calls for research and projects in BRICS,
    272–73
Joint Commission of Taiwan (JCT), 225, 226,
    235n9
Joint Institute for Nuclear Research
    (Russia), 273
jokes, 117, 118, 120

Kalak, Line, 134
Kaleidos Center for Interdisciplinary
    Ethnography (Ecuador), 167–69
Kampo medicine, 227

Kaşdoğan, Duygu, 155, 156, 164, 165

Kaufman, Sharon, 224

Kaweskar National Reserve (Chile), 245

Kenya, coloniality in academia, 162–64

Khadi and Village Industries Commission (India), 42, 55

Khandekar, Aalok, 155–56, 157, 158–59

Kiswahili (language), 174n3

Kleinman, Daniel Lee, 93, 96, 100, 102

Klima-og Miljødepartement (Ministry of Climate and the Environment) (Norway), 149

Knorr, Karin, 125

knowing practices, 132. *See also* infrastructure of knowledge

knowledge: academic, 261, 263, 267–68, 272, 275; centralization of, 2, 21; circulation, 263, 277; commodification, 67; decentralizing, 187–89; extraction from margins, 36, 113; gaps, 95–96, 104–5; hierarchies, 261, 263; hoarding through patents, 49. *See also* infrastructure of knowledge; patents

knowledge-making communities, 115, 116–17, 119

knowledge production, 263, 270, 277; centralization, 3, 5–6, 15, 17, 18; diversification, 4–5, 13, 22n2; standardized forms, 116–17; and validation, 39–44

knowledge production and epistemic authority: challenging established practices, 121–22; classifications, building trust outside, 119–20; epistemic authority, construction of, 120–21; epistemic practices, overlapping, 114–16; epistemic practices at the margins, 116–19; epistemic privilege, 112–13; exclusionary practices, 121; imperial and colonial epistemic privilege in global academia, 113–14; middle voice, 110–11; traditional academic writing and narrative strategies, 122–25

knowledge production from margins, 97–99, 109–10; beekeeper knowledge making, 93–94, 96, 99–102, 103

knowledges contrasted, 143–44

knowledge space, therapeutic space as: concept of, 223; crafting decentralized profession and its limits, 230–33; decentralizing biomedicine in hospice practice, 226–30; hospice wards in Taiwan's health care system, 224–26; in Taiwan's palliative care, 221–24

Korean medicine, 227

Koskinen, Inkeri, 40

Kourany, Janet, 95

Kuhn, Thomas, 51

Kung Tai Catholic Sanipax Socio-Medical Service and Education Foundation (Taiwan), 225

Kuo, Wen-Hua, 76

labor shortages, 202

Laely, Thomas, 79, 85n4

Lakatos, Imre, 51

land use, 147

languages: and colonialism, 48; and infrastructures of knowing, 134, 138–39, 143. *See also* mistranslation; translation

Large Synoptic Survey Telescope. *See* Vera C. Rubin Observatory

Last Hope (Chile), 243

Latin America: academia in, 175n11; agroecological alternatives as epistemic decentralization in, 191–93; neocolonial epistemic politics in, 186; solidarity economy in, 187

Latour, Bruno, 1

lavender, 228

Law, John, 73–74, 134, 223

Law of Natural Protected Areas in National Parks (Argentina), 257n5

legal issues in hospices, 236n19

legal struggles in Sámi resistance, 146–50

Lehuedé, Sebastián, 80–81

Lei Orgânica de Assistência Social (LOAS) (Brazil), 188

lesbian and gay movement, 98

Lickan Antay people, 81–82

Lien Foundation (Singapore), 221

Livanov, Dmitry, 266

Liyong, Taban lo, 169–70

loans, 194

Lobo, Natália, 76

local ecological knowledge, 134–38

local knowledge, 115; bearers (*árbečeahpit*), 133, 134–40, 143, **145**; traditions, 74

loddejeaddjit (people knowledgeable about the *lodden*)135, 137, 149

Lodden Committee Report, 144, 148–50

salmon farming: contamination due to, 244–46; Salmonleaks case, 245

salmon fishing, 147

"salvage" forms of knowledge projects, 45–46

Salvatore, Ricardo D., 76–77

Sámi herders, 146

Sámi knowledge bearers (*árbečeahpit*), 133, 135–36, 137, 143, 144

Sámi language, 134, 136, 138–39, *145*

Sámi people, 131, 134–38, 140, 144, 148

Sámi resistance: colonial history, 133–34; dominant versus precarious knowing practices, 132–33; heterogeneous resources of knowing, 131–32; *lodden* (Sámi spring-winter duck hunt), 134–38; Loddenut-valget (Lodden Committee), 138–44; Lodden Committee Report, 148–50

Sandoval, Chela, 125

Sankey, Kyla, 33

Santos, Boaventura de Sousa, 12, 13–14, 51–54

Saunders, Cicely, 225

scholarship, 155; academia in Latin America, 175n11; on decolonizing university, 153–54; elite, 153–54; by individual institutions, 157; open access initiatives, 161

school meals, 197

Schultz, Eugene, 43

science, technology, and innovation (STI) policy, 253–54, 263, 268; BRICS collaboration, 264, 269–74, 282n5; BRICS Framework Programme, 270, 272; BRICS Funding Working Group, 270, 272; BRICS MoU on STI collaboration, 270–71; BRICS joint calls for research and projects, 272–73; BRICS thematic working groups, 271–72; joint use of research infrastructure in BRICS, 273–74

science, technology, and medicine, 115

science, technology, and society (STS) studies, 166, 168, 176n18, 223, 232

Science and Technology Ministry (Argentina), 252

science policy, 263

Sciences from Below (Harding), 93

scientific capacities, 248–50, 257n6

scientific concerns, socioterritorial turn in, 251–52

Scientific Information Program (Chile), 79

Scientific/Intellectual Movements (SIMS), 40

Scopus index, 157

seed production, 39–40

selective ignorance, 95

Sempreviva Organização Feminista (SOF) (Brazil), 200

sentipensante (feeling-thinking) epistemology, 189

shared decision-making, 33

shared narratives, 137

sharing intelligence, 135

shenshu (accupuncture point), 229

Shibutani, Tamotsu, 119

shiunkou (herbal ointment), 228, 236n14

Shiva, Vandana, 39, 42

short supply chains, 187–88, 198, 199, 200, 201, 203, 204

shushi huli (nursing for comfort), 223, 231–32, 235n4, 236n13

Silicon Savannah (Nairobi), 156, 162

Silva, Luiz Inácio Lula da, 204, 205

Sistema Participativo de Garantia (SPG) (Brazil), 196, 197

Sitas, Ari, 263

small wounds, treating, 236n14

social justice, 159

social movements, 2, 9, 11, 196

social organization of trust, 74

social sciences, 162, 175n13

social skills, 135; sociability, 136

social technology, concept of, 198

Society for Social Studies of Science, 116, 166

socioterritorial turn in scientific concerns, 251–52

solidarity economy, agrifood, 187–89, 195–205

Solidarity Kitchens program (Brazil), 205

Sørensen, Knut, 121, 125

South Africa. *See* BRICS (Brazil, Russia, India, China, and South Africa)

South-South cooperation, 262, 275, 278

SPARC (Scheme for Promotion of Academic Research and Collaboration) (India), 175n8

spiritual consultation, 236n13

spiritual master, 229

Springer (publisher), 162, 175n13

state-led agenda, 278–80

state-led strategy of decentralization: BRICS Network University, document analysis, 265–69; BRICS STI collaboration, document analysis, 269–74; empirical approach, 264–65; example of decentralization, 263–64; overview, 261–63; results, material and symbolic decentralization, 274–80

Statens Naturoppsyn (Nature Inspectorate) (Norway), 135

STI. *See* science, technology, and innovation (STI) policy

Stone, Alluquere, 125

storytelling modes, 147

Strauss, Anselm, 227

Su, Wen-Hao, Dr., 225

Suarez, Maka, 156, 160–62, 167–69

subject formation, 124

subjective and narrative forms, 110–11

Subpesca (Undersecretariat of Fisheries) (Chile), 244

subsidies, 194

subverting epistemic authority, 109–10

Sugawara, Hirotaka, 125

suq'akuna (the dead), 48

surveillance, 121

Suryanarayanan, Sainath, 93, 96, 100, 102

symbolic decentralization: challenge of, 277–78; and material, 274–80

Taiwan: health care system, hospice wards in, 224–26; palliative care, invisible knowledge spaces in, 221–24

Taiwan Academy of Hospice Palliative Medicine (TAHPM), 225–26, 231

Taiwan Association for Medical Continuing Education and Promotion, 236n18

Taiwan Association of Hospice Palliative Nursing, 231

Taiwan Community Hospital Association, 236n18

Taiwan Hospital Association, 235n9

Taiwan Medical Association, 235n6, 235n9

Taiwan Nongovernmental Hospitals and Clinics Association, 235n9

Taiwan Society of Cancer Palliative Medicine, 236n18

Tanio, Nadine, 125

Tapuya: Latin American Science, Technology and Society (journal), 166, 84n3

Taylor and Francis (publisher), 161, 162, 175n12, 175n13

tea tree, 228

technical skills in agroecological production, 200

Technology Innovation Agency (South Africa), 283n10

templates, 109–10, 120, 123

Teresa Lozano Long Institute of Latin American Studies (LLILAS), 68

Terram Foundation (Chile), 243, 245

territorial development policies (PRONAT) (Brazil), 202

testimonial injustice, 1, 10

textual knowing, 139–43, 145

Tharoor, Shashi, 266

thematic working groups on STI (BRICS) 271–72

theory from the South, 10

therapeutic space as knowledge space. *see* knowledge space, therapeutic space as

therapeutic touch, 236n12

Tiirasaari Island (Finland), 146

Tousignant, Noémi, 164

toxicologists, 99–100, 164

traditional ecological knowledge (TEK), 134–35, 142

traditional foods, 195

traditional knowledges, 115. *See also* local knowledge

traditional skills, transferring, 147

training courses for physicians and nurses, 236n19

transcutaneous electrical nerve stimulator (TENS), 228–29, 231, 236n15

transgressive knowledge, 124–25

translation, 48–49, 136, 138–39

transnational alliance, 262, 271, 276, 279, 280

transnational corporate entities, 33, 39

transnationalizing STS (science, technology, and society), 116

transnational pharmaceutical companies, 39

transnational WGFS (women's, gender, feminist, and sexuality), 116

Traweek, Sharon, 154

treatment activists, 97–99

www.ingramcontent.com/pod-product-compliance
Lightning Source LLC
Chambersburg PA
CBHW020825270326
41928CB00006B/443